T0320686

Attractors, Shadowing, and Approximation of Abstract Semilinear Differential Equations

Other World Scientific Titles by the Author

Lectures in Nonlinear Functional Analysis: Synopsis of Lectures Given at the Faculty of Physics of Lomonosov Moscow State University
ISBN: 978-981-124-892-4

ATTRACTORS, SHADOWING, AND APPROXIMATION OF ABSTRACT SEMILINEAR DIFFERENTIAL EQUATIONS

Sergey I Piskarev

M V Lomonosov Moscow State University, Russia & Mari State University, Russia &
Bauman Moscow State Technical University, Russia &
Russian Institute for Scientific and Technical Information, Russia

Alexey V Ovchinnikov

M V Lomonosov Moscow State University, Russia &
Russian Institute for Scientific and Technical Information, Russia

 **World Scientific**

N JERSEY · LONDON · SINGAPORE · BEIJING · SHANGHAI · HONG KONG · TAIPEI · CHENNAI · TOKYO

Published by

World Scientific Publishing Co. Pte. Ltd.

5 Toh Tuck Link, Singapore 596224

USA office: 27 Warren Street, Suite 401-402, Hackensack, NJ 07601

UK office: 57 Shelton Street, Covent Garden, London WC2H 9HE

British Library Cataloguing-in-Publication Data
A catalogue record for this book is available from the British Library.

ISBN 978-981-127-277-6 (hardcover)
ISBN 978-981-127-278-3 (ebook for institutions)
ISBN 978-981-127-279-0 (ebook for individuals)

For any available supplementary material, please visit
https://www.worldscientific.com/worldscibooks/10.1142/13320#t=suppl

Printed in Singapore

Preface

The book is devoted to some branches of the theory of approximation of abstract differential equations, namely, approximation of attractors in the case of hyperbolic equilibrium points, shadowing, and approximation of time-fractional semilinear problems. We discuss and examine such concepts as abstract parabolic equations, general approximation schemes, compact convergence, attractors, unstable and stable manifolds, upper and lower semicontinuity of attractors, affinity principle, principle of compact approximation, semilinear differential equations in Banach spaces, periodic solutions of semilinear equations, stability of solution in Lyapunov's sense, hyperbolic equilibrium points, semiflows, rotation of vector fields, index of a solution, shadowing, analytic C_0-semigroups, Banach spaces, semidiscretization, discretization in space and in time, fractional equations, fractional powers of operators, condensing operators, etc. In this monograph, the most important and principal results scattered in many research papers are collected together and systemized. The book can be useful for specialists in partial differential equations, functional analysis, theory of approximation of differential equations, and for all researchers, students, and postgraduates who apply these branches of mathematics in their work.

Contents

Introduction

In this survey, we present the basic theory and more interesting developments of resolution families for the abstract semilinear Cauchy problem in a Banach space E

$$\frac{du(t)}{dt} = Au(t) + f(t, u(t)), \quad 0 \le t < T, \quad u(0) = u^0, \qquad (*)$$

where A generates an exponentially decaying, compact analytic C_0-semigroup in the Banach space E and $f(\cdot, \cdot)$ is a globally Lipschitzian and bounded map from $[0, T] \times E^\theta$ into E ($E^\theta = D((-A)^\theta)$ with the graph norm), which recently appeared in the literature (see [68, 72, 105, 125, 129, 169, 222, 227, 260, 269, 301]).

For the dynamical system

$$u'(t) = Au(t) + f(u(t)), \quad u(0) = u^0, \qquad (**)$$

it was shown that if the abstract parabolic problem has only hyperbolic equilibria, then all equilibrium points are isolated and there is only a finite number of them. Moreover, the number of such equilibrium solutions is odd. The main assumptions are the compactness of the resolvent, the hyperbolicity of any equilibrium point, and the uniform boundedness of the nonlinear function $f(\cdot)$. In addition, if, in the case where the system has gradient structure, we assume that every global solution converges as $t \to \pm\infty$, then it has a global attractor, which is the union of unstable manifolds of equilibrium points. We allow rather general discretization schemes following the theory of discrete approximations developed by F. Stummel, R. D. Grigorieff, and G. Vainikko. Under such assumptions, there is a possibility of considering approximations of attractors on rather general approximation scheme that include finite-element, projection, and finite-difference methods. Then we prove that within these approximation schemes, attractors also behave lower semicontinuously.

In general, the equiattraction of global attractors $\mathcal{A}_n$ of the dynamical systems $S_n(\cdot)$ in Banach spaces $(E_n, \|\cdot\|)$ as $n \to \infty$ is shown to be equivalent to the continuous convergence of $\mathcal{A}_n$ defined in terms of the concept of the discrete convergence. The results are applied to general approximation schemes for abstract semilinear parabolic problems of the form $(**)$, where A generates an exponentially decaying, compact, analytic C_0-semigroup in a Banach space E and $f(\cdot) : E^\theta \to E$ is globally Lipschitzian and bounded.

One can develop a general approach to establish a discrete dichotomy in a very general setting and prove shadowing theorems that compare solutions of the continuous problem with solutions of discrete approximations in space and time. In [56], a space discretization was constructed under the assumption that the resolvent is compact. It is well known (see [54, 173, 175, 176]) that the phase space in a neighborhood of a hyperbolic equilibrium can be split in a such way that the original initial-value problem is reduced to initial-value problems with exponential bounded solutions on the corresponding subspaces. We show that such a decomposition of the flow persists under rather general approximation schemes based on the property of uniform condensing. The main assumption of our results are naturally satisfied, in particular, for operators with compact resolvents and condensing semigroups (see [20, 33]).

Acknowledgments

This work was supported by the Russian Science Foundation (projects No. 20-11-20085 for Chapter 5 and No. 23-21-00005 for Chapter 6).

Preliminaries

The behavior of smooth dynamical system in a neighborhood of an equilibrium point can be mostly determined by the behavior of the linearized system in this neighborhood. In this connection, the concept of stability of a solution is very important for the Cauchy problem $u'(t) = Au(t)$, $u(0) = u^0$, with a linear operator A in a Banach space E.

Definition 1. A C_0-semigroup $\exp(tA)$, $t \geq 0$, on a Banach space E is called

(a) *uniformly exponentially stable* if there exists $\mu > 0$ such that
$$\lim_{t \to \infty} e^{\mu t} \| \exp(tA) \| = 0;$$

(b) *uniformly stable* if
$$\lim_{t \to \infty} \| \exp(tA) \| = 0;$$

(c) *strongly stable* if
$$\lim_{t \to \infty} \| \exp(tA)x \| = 0 \quad \text{for any } x \in E; \tag{$\star$}$$

(d) *weakly stable* if
$$\lim_{t \to \infty} \langle \exp(tA)x, x^* \rangle = 0 \quad \text{for any } x \in E, \ x^* \in E^*.$$

The most important facts in this direction were obtained in [106].

Theorem 1. *A C_0-semigroup $\exp(tA)$, $t \geq 0$, on a Hilbert space H is uniformly exponentially stable if and only if the half-plane $\operatorname{Re}\lambda > 0$ is in the resolvent set $\rho(A)$ and*
$$C = \sup_{\operatorname{Re}\lambda > 0} \| (\lambda I - A)^{-1} \| < \infty.$$

The thorough explanation of the properties introduced in Definition 1 is given in [93]. We can also note the following result (see [20, 302]).

Theorem 2. *Assume that a uniformly bounded C_0-semigroup $\exp(tA)$, $t \geq 0$, on a Banach space E is such that the following conditions are fulfilled:*

(i) $P\sigma(A^*) \cap i\mathbb{R} = \varnothing$;
(ii) $\sigma(A) \cap i\mathbb{R}$ *is countable.*

Then the C_0-semigroup $\exp(tA)$, $t \geq 0$, is strongly stable, i.e.,

$$\lim_{t \to \infty} \exp(tA)x = 0 \quad \text{for all } x \in E.$$

The strong stability of C_0-semigroup were studied in many papers (see, e.g., [46, 50, 224, 276]). We mention the following result.

Theorem 3 (see [46]). *Assume that a C_0-semigroup $\exp(tA)$ acting on a Banach space E is slowly varying on infinity and $x_0 \in E$. Then the limit of $\exp(tA)x_0$ as $t \to \infty$ exists if and only if the limit*

$$\lim_{0 < \epsilon \to 0} \epsilon(\epsilon I - A)^{-1}x_0$$

exists. In this case,

$$\lim_{t \to \infty} \exp(tA)x_0 = \lim_{0 < \epsilon \to 0} \epsilon(\epsilon I - A)^{-1}x_0 = y_0.$$

Moreover, the element y_0 belongs to the null space $\mathcal{N}(A)$. The semigroup $\exp(\cdot A)$ is strongly stable (i.e., $(\star)$ holds) if and only if

$$\lim_{0 < \epsilon \to 0} \epsilon(\epsilon I - A)^{-1}x = 0 \quad \text{for any } x \in E.$$

We mention the following interesting fact: resolution families $S_\alpha(t, A)$, $t \geq 0$, for fractional equations of order $0 < \alpha < 1$ never decrease exponentially (see [189]). So, it is very important to study the case of just strongly stable resolution families. For the asymptotic property, see also [312].

We know (see [241]) that the α-resolvent family is a special case of families of resolvents arising from the Volterra integral equation

$$S_\alpha(t)x = x + \int_0^t a(t - s)AS_\alpha(s)x\,ds$$

with $a(t) = t^{\alpha - 1}/\Gamma(\alpha)$. Note that for an analytic α-resolvent family $S_\alpha(t)$, if there exists $\delta < 0$ such that $1/\hat{a}(z) \in \rho(A)$ and the Laplace transform of $S_\alpha(t)$, which is equal to $\dfrac{(1/\hat{a}(z) - A)^{-1}}{z\hat{a}(z)}$, is analytic for $\operatorname{Re} z > \delta$, then $S_\alpha(t)$ can be exponentially stable (see [97]). However, the function $a(\cdot)$, has the form of a function of exponential type (see also [273]).

Chapter 1

Semilinear Equations

The analysis of semidiscretizations and discretization in time of Cauchy problems under the compactness condition for resolvents for the first- and second-order equations was developed in [95, 233, 235]. In this chapter, we present the analysis of semidiscretization of periodic problems and Cauchy problems for first-order equations in Banach spaces.

1.1 Differential Equations in Banach Spaces

Let $B(E)$ be a Banach algebra of all linear bounded operators on a complex Banach space E. We denote by $\mathcal{C}(E)$ the set of all linear, closed, densely defined operators on E, by $\sigma(B)$ the spectrum of an operator B, by $\rho(B)$ the resolvent set of B, by $\mathcal{N}(B)$ the null-space of B, and by $\mathcal{R}(B)$ the range of B. Recall that an operator $B \in B(E)$ is called a *Fredholm operator* if $\mathcal{R}(B)$ is closed, $\dim \mathcal{N}(B) < \infty$, and $\operatorname{codim} \mathcal{R}(B) < \infty$. The index of B is the number $\operatorname{ind}(B) = \dim \mathcal{N}(B) - \operatorname{codim} \mathcal{R}(B)$.

In a Banach space E, let us consider the semilinear Cauchy problem

$$v'(t) = Av(t) + f\big(t, v(t)\big), \quad v(0) = v^0, \quad t \in \overline{\mathbb{R}}^+ = [0, \infty), \qquad (1.1.1)$$

and also the T-periodic problem

$$u'(t) = Au(t) + f\big(t, u(t)\big), \quad u(t) = u(T + t), \quad t \in \overline{\mathbb{R}}^+, \qquad (1.1.2)$$

with an operator A generating an analytic, compact C_0-semigroup and a sufficiently smooth function $f(\cdot, \cdot)$ satisfying the condition

$$f(t, x) = f(t + T, x) \quad \text{for any } x \in E \text{ and } t \in \overline{\mathbb{R}}^+.$$

We treat solutions in the classical sense (see [295]). The existence of solutions for such problems was proved, for example, in [32, 38, 58, 72, 125, 129,

$130, 168, 215, 216]$. Here and below, $A : D(A) \subseteq E \to E$ is a closed linear operator such that

$$\|(\lambda I - A)^{-1}\|_{B(E)} \le \frac{M}{1 + |\lambda|} \quad \text{for any } \operatorname{Re} \lambda \ge 0. \tag{1.1.3}$$

Under the condition (1.1.3) the spectrum of A lies in the left half-plane: $\sup\{\operatorname{Re} \lambda : \lambda \in \sigma(A)\} < 0$, so one can define the fractional-power operator $(-A)^\theta$, $\theta \in \mathbb{R}^+$, associated with A and E^θ (see $[124, 129, 170, 207]$), where $E^\theta := D((-A)^\theta)$ endowed with the graph norm $\|x\|_{E^\theta} = \|(-A)^\theta x\|_E$.

Denote by $v(\cdot; v^0)$ the solution of the Cauchy problem (1.1.1) with the initial data $v(0; v^0) = v^0$. This function $v(\cdot; v^0)$ satisfies the integral equation

$$v(t) = \exp(tA)v^0 + \int_0^t \exp\big((t - s)A\big)f\big(s, v(s)\big)ds, \quad t \in [0, T]. \tag{1.1.4}$$

Then one can introduce the shift operator $\mathcal{K}(v^0) = v(T; v^0)$, which maps E into E. If $v(\cdot; x^*)$ is a periodic solution of (1.1.1), then x^* is the zero of the compact vector field determined by $I - \mathcal{K}$, i.e., $\mathcal{K}(x^*) = x^*$.

Let Ω be an open set in a Banach space F and $\mathcal{B} : \bar{\Omega} \to F$ be a compact operator having no fixed points on the boundary of Ω. Then for the vector field $\mathcal{V}(x) = x - \mathcal{B}x$, the rotation of the vector field $\gamma(I - \mathcal{B}; \partial\Omega)$ is defined, which is an integer-valued characteristic of this field (see $[168]$). Let z^* be a unique isolated fixed point of the operator $\mathcal{B}$ in the ball $\mathcal{U}_F(z^*, r_0)$ of radius r_0 centered at z^*. Then

$$\gamma\big(I - \mathcal{B}; \partial\mathcal{U}_F(z^*, r_0)\big) = \gamma\big(I - \mathcal{B}; \partial\mathcal{U}_F(z^*, r)\big)$$

for $0 < r \le r_0$, and this common value of the rotations is called the *index* of the fixed point z^* and is denoted by $\operatorname{ind}(z^*; I - \mathcal{B})$.

Note that the results concerning problems similar to (1.1.1) in the case where $\exp(\cdot A)$ is not compact but is a condensing semigroup can be found, for example, in $[20, 66, 215, 216]$.

We also define an operator K such that the solution of (1.1.2) satisfies the integral equation

$$u(t) = (Ku)(t) \equiv \exp(tA)\Big(I - \exp(TA)\Big)^{-1}$$
$$\times \int_0^T \exp\big((T - s)A\big)f\big(s, u(s)\big)ds$$
$$+ \int_0^t \exp\big((t - s)A\big)f\big(s, u(s)\big)ds, \tag{1.1.5}$$

where T is the period of a solution $u(\cdot)$. The operator K maps the space of continuous functions $C([0,T];E)$ into itself.

Remark 1.1. In (1.1.5), we assume that the operator $\left(I - \exp(TA)\right)^{-1}$ exists and is bounded. Meanwhile, it suffices to assume that there exists $t_0 > 0$ such that $\left(I - \exp(tA)\right)^{-1} \in B(E)$ for $t \geq t_0$. This assumption is not restrictive since, without loss of generality, we are able to replace A by $A - \omega I$ and obtain

$$\left\| \exp\left(t(A - \omega)\right) \right\| \leq M e^{-\delta t}, \quad \delta > 0, \quad t \geq 0.$$

By [41], we have

$$\left(I - \exp(tA)\right)^{-1} \in B(E), \quad t > 0.$$

Remark 1.2. We use the phrase *"a function $f(\cdot,\cdot)$ is sufficiently smooth"* in the following sense: $f(\cdot,\cdot)$ is at least continuous in both arguments,

$$\sup_{\substack{t \in [0,T], \\ \|x\| \leq c_1}} \left\| f(t,x) \right\| \leq C_2,$$

and it is such that a global mild solution of the problem (1.1.1) exists.

1.2 Some Interesting Properties

Theorem 1.1 (the principle of affinity; see [57]). *Assume that a function $f(\cdot,\cdot)$ is sufficiently smooth, so that there exists an isolated mild solution $u^*(\cdot)$ of the periodic problem (1.1.2). In addition, let an element x^* be a fixed point of $\mathcal{K}$, and let the resolvent $(\lambda I - A)^{-1}$ be compact. Then the following equality holds:*

$$\operatorname{ind}(x^*; I - \mathcal{K}) = \operatorname{ind}(u^*(\cdot); I - K).$$

Proof. Let $S \equiv \mathcal{U}_E(x^*, \rho)$. Then the rotation $\gamma(I - \mathcal{K}; \partial S)$ of the field $I - \mathcal{K}$ on the sphere ∂S is equal to the index $\operatorname{ind}(x^*; I - \mathcal{K})$:

$$\gamma(I - \mathcal{K}; \partial S) = \operatorname{ind}(x^*; I - \mathcal{K}). \tag{1.2.1}$$

Recall that we consider the operator K defined by (1.1.5) on the space

$$F = C([0,T];E) \equiv \left\{ u(t) : \|u\|_F = \max_{t \in [0,T]} \|u(t)\|_E < \infty \right\}.$$

Let

$$M = \sup_{x \in S} \max_{0 \leq t \leq T} \|v(t;x)\|. \tag{1.2.2}$$

Consider the domain

$$\Omega = \Big\{ u(\cdot) \in C([0,T];E) : \ u(0) \in S, \ \|u(\cdot)\|_F \le M+1 \Big\} \subset C([0,T];E).$$

The function $u^*(\cdot)$ is a unique zero of the compact vector field $I - K$ on $\overline{\Omega}$. Therefore,

$$\gamma(I - K; \partial\Omega) = \operatorname{ind}(u^*(\cdot); I - K). \tag{1.2.3}$$

By virtue of (1.2.1) and (1.2.3), to complete the proof of the affinity theorem it suffices to show that

$$\gamma(I - \mathcal{K}; \partial S) = \gamma(I - K; \partial\Omega). \tag{1.2.4}$$

To do this, we consider the following family of compact vector fields on $\partial\Omega$:

$$\begin{aligned}
\Phi\big(u(\cdot); \lambda\big) = u(t) &- (1 - \lambda) \exp(tA)\Big(I - \exp(TA)\Big)^{-1} \\
&\times \int_0^T \exp\big((T - s)A\big) f\big(s; u(s)\big) ds - \lambda \exp(tA)\mathcal{K}\big(u(0)\big) \\
&- \int_0^t \exp\big((t - s)A\big) f\big(s, u(s)\big) ds, \quad 0 \le \lambda \le 1.
\end{aligned} \tag{1.2.5}$$

The fields $\Phi\big(u(\cdot); \lambda\big)$ are nondegenerate on $\partial\Omega$. Indeed, if we have $\Phi\big(u_0(\cdot); \lambda_0\big) = 0$ for some $u_0(\cdot) \in \partial\Omega$ and $\lambda_0 \in [0, 1]$, then

$$\begin{aligned}
u_0(0) = (1 - \lambda_0)\Big(I &- \exp(TA)\Big)^{-1} \\
&\times \int_0^T \exp\big((T - s)A\big) f\big(s, u_0(s)\big) ds + \lambda_0 u_0(T).
\end{aligned} \tag{1.2.6}$$

Since Eq. (1.2.6) and the relation $\Phi\big(u_0(\cdot); \lambda_0\big) = 0$ imply that the function $u_0(\cdot)$ is a mild solution of (1.1.1), we obtain

$$\int_0^T \exp\big((T - s)A\big) f\big(s, u_0(s)\big) ds = u_0(T) - \exp(TA)u_0(0). \tag{1.2.7}$$

Without loss of generality, we can assume that $\operatorname{Re}\sigma(A) < 0$. From (1.2.6) and (1.2.7) we obtain

$$\exp(TA)\big((u_0(0) - u_0(T)\big) = \lambda_0^{-1}\big(u_0(0) - u_0(T)\big).$$

If $u_0(0) - u_0(T) \ne 0$, then this element is an eigenvector of the operator $\exp(TA)$ with eigenvalue $\lambda_0^{-1} > 1$. This is impossible since

$$\operatorname{Re}\sigma(A) < 0 \quad \text{and} \quad \sigma\big(\exp(TA)\big) \setminus \{0\} = e^{T\sigma(A)}.$$

Therefore, $u_0(0) = u_0(T)$, so that $u_0(\cdot)$ is a T-periodic solution of the problem (1.1.2) and it is a zero of the field $I - K$; a contradiction.

The fields of the family (1.2.5) are nondegenerate on $\partial\Omega$; therefore, the fields $\Phi(u(\cdot); 0) = I - K$ and $\Phi(u(\cdot); 1)$ are homotopic on $\partial\Omega$. We obtain

$$\gamma(I - K; \partial\Omega) = \gamma\Big(\Phi(u(\cdot); 1); \partial\Omega\Big). \tag{1.2.8}$$

Consider on $\partial\Omega$ the following family of vector fields:

$$\Psi(u(\cdot); \lambda) = u(t) - P_\lambda\Big(\exp(tA)\mathcal{K}(u(0)) + \int_0^t \exp\big((t - s)A\big)f\big(s, u(s)\big)ds\Big), \tag{1.2.9}$$

where $0 \leq \lambda \leq T$ and the operator $P_\lambda : F \to F$ is defined as follows:

$$(P_\lambda w)(t) = \begin{cases} w(t) & \text{for } 0 \leq t \leq \lambda, \\ w(\lambda) & \text{for } \lambda \leq t \leq T. \end{cases}$$

The operator

$$Q(u(\cdot))(t) = \exp(tA)\mathcal{K}(u(0)) + \int_0^t \exp\big((t - s)A\big)f\big(s, u(s)\big)ds,$$

which maps F to F, is compact. The operator $P_\lambda : F \to F$ is strongly continuous in λ. Therefore, the operator $P_\lambda Q$ is uniformly continuous in λ and hence the family (1.2.9) is a compact connection (see [168, Sec. 19.1]).

Let us show that the family (1.2.9) is nondegenerate on $\partial\Omega$. Assume that for some $\lambda_0 \in [0, 1]$ and $u_0(\cdot) \in \partial\Omega$, we have

$$u_0(\cdot) \neq v(\cdot; x^*), \quad \Psi(u_0(\cdot); \lambda_0) = 0.$$

The boundary $\partial\Omega$ of the domain Ω consists of two sets:

$$G_0 = \Big\{u(\cdot) \in C([0, T]; E) : u(0) \in S, \|u(\cdot)\|_{C([0,T];E)} = M + 1\Big\},$$

$$G_1 = \Big\{u(\cdot) \in C([0, T]; E) : u(0) \in \partial S, \|u(\cdot)\|_{C([0,T];E)} \leq M + 1\Big\}.$$

Let $u_0(\cdot) \in G_0$. Then

$$\|u_0(\cdot)\|_{C([0,T];E)} = M + 1. \tag{1.2.10}$$

On the other hand, since the function $u_0(\cdot)$ is a solution of (1.1.1) on the segment $[0, \lambda]$ and $u_0(0) \in S$, it follows from (1.2.2) that

$$\|u_0(\cdot)\|_{C([0,T];E)} = \max_{0 \leq t \leq T} \|u_0(t)\|_E = \max_{0 \leq t \leq \lambda} \|u_0(t)\|_E \leq M. \tag{1.2.11}$$

Equalities (1.2.10) and (1.2.11) are incompatible. So we have the only one possibility $u_0(\cdot) \in G_1$ and $u_0(0) \in \partial S$. The relation $\Psi(u_0(\cdot); \lambda_0) = 0$ implies that $u_0(0) = \mathcal{K}(u_0(0))$, which is impossible due to the choice of the radius ρ of the ball S. Therefore, the fields of the family (1.2.9) are nondegenerate on $\partial\Omega$ and homotopic to each other. So,

$$\gamma\Big(\Psi\big(u(\cdot);0\big);\partial\Omega\Big) = \gamma\Big(\Psi\big(u(\cdot);T\big);\partial\Omega\Big).$$

However, $\Psi\big(u(\cdot);T\big) = \Phi\big(u(\cdot);1\big)$; therefore,

$$\gamma\Big(\Psi\big(u(\cdot);0\big);\partial\Omega\Big) = \gamma\Big(\Phi\big(u(\cdot);1\big);\partial\Omega\Big). \tag{1.2.12}$$

Consider the field

$$\Psi\big(u(\cdot);0\big) = u(t) - \mathcal{K}\big(u(0)\big), \quad u(\cdot) \in \partial\Omega.$$

Since the operator $\mathcal{K}\big(u(0)\big)$ can also be considered as a mapping from F to the space $\widetilde{E}$ of constant functions, we conclude (see [168]) that its rotation coincides with the rotation of its restriction $\widetilde{\Psi}$ to $\partial\Omega \cap \widetilde{E}$ and with the rotation of its restriction $\widetilde{\Psi}\big(u(\cdot);0\big)$ to $\partial\Omega \cap \widetilde{E}$. Since the field $\widetilde{\Psi}\big(u(\cdot);0\big)$ on $\partial\Omega \cap \widetilde{E}$ is isomorphic to $I - \mathcal{K}$ on ∂S, we have

$$\gamma\Big(\Psi\big(u(\cdot);0\big);\partial\Omega\Big) = \gamma\Big(\widetilde{\Psi}\big(u(\cdot);0\big);\partial\Omega \cap \widetilde{E}\Big) = \gamma(I - \mathcal{K};\partial S). \tag{1.2.13}$$

From Eqs. (1.2.8), (1.2.12), and (1.2.13) we obtain (1.2.4). The theorem is proved. □

Remark 1.3. To prove the compactness of the operator

$$\int_0^t \exp\big((t-s)A\big)f\big(s,u(s)\big)ds,$$

one usually uses a semigroup property. In the case of another strongly continuous compact function instead of $\exp(\cdot A)$ (for example, the sine operator function), one can apply an approach from [300]. For the case of fractional differential equations $D^\alpha u(t) = Au(t) + f(u(t))$, one can state the compactness of the operator for general formula

$$u(t) = S_\alpha(t)u^0 + \int_0^t P_\alpha(t-s)f(u(s))ds$$

using an approach from [96].

Let us consider in the space E the following semilinear, nonautonomous parabolic problem:

$$\begin{aligned} u'(t) &= Au(t) + f(t,u(t)), \quad t \in \overline{\mathbb{R}}^+, \\ u(0) &= u^0 \in D(A), \end{aligned} \tag{1.2.14}$$

where A generates an analytic C_0-semigroup and $f(\cdot, \cdot)$ is sufficiently smooth. A classical solution of (1.2.14) satisfies the equation

$$u(t) = \tilde{K}u(t) \equiv \exp(tA)u^0 + \int_0^t \exp\big((t-s)A\big)f\big(s, u(s)\big)ds, \quad t \in \overline{\mathbb{R}}^+,$$
$$(1.2.15)$$

where $\tilde{K}$ is treated as an operator from F to F. We will need the following result.

Theorem 1.2 (see [121]). *Let A be the generator of an analytic C_0-semigroup, the resolvent $(\lambda I - A)^{-1}$ be compact for a certain $\lambda \in \rho(A)$, and let the operator $\tilde{K}$ be given by the formula (1.2.15). If $u^*(\cdot)$ is a unique mild solution of the problem (1.2.14), then $\text{ind}(u^*(\cdot); I - \tilde{K}) = 1$.*

If A is as above and $0 < \theta < 1$ is fixed, consider the following semilinear, autonomous parabolic problem:

$$u'(t) = Au(t) + f(u(t)), \quad t \in \overline{\mathbb{R}}^+,$$
$$u(0) = u^0 \in E^\theta,$$
$$(1.2.16)$$

in the space E^θ, where $f(\cdot) : E^\theta \subseteq E \to E$ is a globally Lipschitzian, bounded, and continuously Fréchet differentiable function.

Note that even in case where

$$u(t) = \exp(tA)u^0 + \int_0^t \exp\big((t-s)A\big)g(s)ds, \quad t \geq 0,$$

and the C_0-semigroup $\exp(\cdot A)$ is analytic, the function $u(\cdot)$ is not necessarily differentiable if $g(\cdot) \in C([0, T]; E)$, i.e.,

$$u'(t) = Au(t) + g(t), \quad t \in [0, T],$$
$$u(0) = u^0,$$
$$(1.2.17)$$

is not classically well posed in $C([0, T]; E)$ for a general Banach space E. However, the problem (1.2.17) is classically well posed in $C([0, T]; E^\theta)$ (see [81, 82, 266]), where $E^\theta = (E, D(A))_\theta$ is a suitable interpolation space. Moreover, if one has in mind applications like the space $E = L^p(\Omega)$ and needs Fréchet differentiability of the function $f(\cdot) : L^2(\Omega) \to L^2(\Omega)$, then the function $f(\cdot)$ must be linear (cf. [11]). To cover more general nonlinearities, one must deal with a weaker assumption that $f(\cdot) : H_0^1(\Omega) \to L^2(\Omega)$ is Fréchet differentiable. The difficulties caused by these facts can be resolved by considering the problem (1.2.16) in a Banach space E^θ, $0 \leq \theta < 1$, and assume that the mapping $f(\cdot) : E^\theta \to E$ is Fréchet differentiable with derivative at the equilibrium $f'(u^*) \in B(E^\theta, E)$. In the case where $E = L^2(\Omega)$ and $A = \Delta$, we normally have $E^{1/2} = H_0^1(\Omega)$ for $\theta = 1/2$. Theorem 1.2 can be proved for the setting (1.2.16).

1.3   Theorem ABC

F. Stummel, R. Grigoriev, and G. Vainikko proposed the idea of discrete convergence or $\mathcal{P}$-*convergence* in the context of what they called a general approximation scheme (see [112–116, 271, 285, 288]), which can be described as follows. Let E_n and E be Banach spaces and $\{p_n\}$ be a sequence of linear bounded operators $p_n : E \to E_n$, $p_n \in B(E, E_n)$, $n \in \mathbb{N} = \{1, 2, \dots\}$, possessing the following property:

$$\|p_n x\|_{E_n} \to \|x\|_E \quad \text{as } n \to \infty \text{ for any } x \in E. \tag{1.3.1}$$

The uniform boundedness principle (see [313]) implies the following assertion.

Lemma 1.1 (see [288]). *There is a constant $C \geq 1$ such that*

$$\|p_n\|_{B(E, E_n)} \leq C \quad \text{for all } n \in \mathbb{N}.$$

Definition 1.1. We say that a sequence $\{x_n\}$ of elements $x_n \in E_n$, $n \in \mathbb{N}$, $\mathcal{P}$-*converges to* $x \in E$ (notation $x_n \xrightarrow{\mathcal{P}} x$) if $\|x_n - p_n x\|_{E_n} \to 0$ as $n \to \infty$.

Definition 1.2. We say that a sequence of bounded linear operators $B_n \in B(E_n)$, $n \in \mathbb{N}$, $\mathcal{PP}$-*converges* to a bounded linear operator $B \in B(E)$ (notation $B_n \xrightarrow{\mathcal{PP}} B$) if $B_n x_n \xrightarrow{\mathcal{P}} Bx$ for any $x \in E$ and any sequence $\{x_n\}$, $x_n \in E_n$, $n \in \mathbb{N}$, such that $x_n \xrightarrow{\mathcal{P}} x$.

Let us consider the following well-posed Cauchy problem in the Banach space E with an operator $A \in \mathcal{C}(E)$:

$$u'(t) = Au(t), \quad t \in \overline{\mathbb{R}}^+, \quad u(0) = u^0, \tag{1.3.2}$$

where the operator A generates a C_0-semigroup $\exp(\cdot A)$. It is well known (see [292]) that this C_0-semigroup gives the solution of (1.3.2) by the formula $u(t) = \exp(tA)u^0$, $t \geq 0$.

On the general discretization scheme, let us consider the following semidiscrete approximation of the problem (1.3.2) in the Banach spaces E_n:

$$u'_n(t) = A_n u_n(t), \quad t \in \overline{\mathbb{R}}^+, \quad u_n(0) = u_n^0, \tag{1.3.3}$$

where the operators $A_n \in \mathcal{C}(E_n)$ generate C_0-semigroups that are compatible with the operator A and $u_n^0 \xrightarrow{\mathcal{P}} u^0$.

We present the following version (only for analytic C_0-semigroups) of the Trotter–Kato theorem, which solves the problem of the convergence of solutions of the problems (1.3.3) to a solution of the problem (1.3.2).

Theorem 1.3 (see [120, 228, 235], Theorem ABC). *Let the operators A and A_n be generators of analytic C_0-semigroups. The following conditions (A) and (B$_1$) together are equivalent to Condition (C$_1$):*

(A) Compatibility: *there exists $\lambda \in \left[\bigcap_n \rho(A_n) \right] \cap \rho(A)$ such that the resolvents converge:*

$$(\lambda I_n - A_n)^{-1} \xrightarrow{\mathcal{PP}} (\lambda I - A)^{-1}.$$

(B$_1$) Stability: *there exist constants $M_2 \geq 1$ and ω_2 such that*

$$\left\| (\lambda I_n - A_n)^{-1} \right\|_{B(E_n)} \leq \frac{M_2}{|\lambda - \omega_2|}, \quad \operatorname{Re} \lambda > \omega_2, \quad n \in \mathbb{N};$$

(C$_1$) Convergence: *for any finite $\mu > 0$ and some $0 < \varphi < \pi/2$, we have*

$$\max_{\eta \in \Sigma(\varphi, \mu)} \left\| \exp(\eta A_n) u_n^0 - p_n \exp(\eta A) u^0 \right\|_{E_n} \to 0$$

as $n \to \infty$ whenever $u_n^0 \xrightarrow{\mathcal{P}} u^0$.

Here

$$\Sigma(\varphi, \mu) = \left\{ z \in \Sigma(\varphi) : |z| \leq \mu \right\}, \quad \Sigma(\varphi) = \left\{ z \in \mathbb{C} : |\arg(z)| \leq \varphi \right\}.$$

So, in the investigation of semidiscretization of the semilinear problem (1.1.2), we will assume without loss of generality that Conditions (A) and (B$_1$) are fulfilled.

Theorem ABC was originally proved for C_0-semigroups. We use the version for analytic semigroups due to the fact that the condition of coercive well-posedness of a Cauchy problem is equivalent to the condition of analyticity of a C_0-semigroup, and we will operate by these assumptions.

1.4 Convergence of Semidiscretizations for Periodic Problem

By a semidiscrete approximation of the T-periodic problem (1.1.2) we mean the following sequence of T-periodic problems ($n \in \mathbb{N}$):

$$u_n'(t) = A_n u_n(t) + f_n\big(t, u_n(t)\big), \quad u_n(t) = u_n(t + T), \quad t \in \overline{\mathbb{R}}^+, \quad (1.4.1)$$

where, for each $n \in \mathbb{N}$, the operators A_n generate analytic semigroups in E_n and Condition (A) is satisfied. In addition, the sequence of functions $f_n(\cdot, \cdot)$ is uniformly bounded in the following sense:

$$\sup_{\substack{t \in [0,T], \\ \|x_n\| \leq c_1}} \|f_n(t, x_n)\| \leq C_2 < \infty,$$

the functions f_n approximating f, are sufficiently smooth and satisfy the condition $f_n(t, x_n) = f_n(t + T, x_n)$ for any $x_n \in E_n$ and $t \in \overline{\mathbb{R}}^+$.

Mild solutions of (1.4.1) are determined by the equations

$$u_n(t) = (K_n u_n)(t)$$

$$\equiv \exp(tA_n)\Big(I_n - \exp(TA_n)\Big)^{-1} \int_0^T \exp\big((T - s)A_n\big) f_n\big(s, u_n(s)\big) ds$$

$$+ \int_0^t \exp\big((t - s)A_n\big) f_n\big(s, u_n(s)\big) ds. \tag{1.4.2}$$

There has been a great interest in these problems in the literature (see, e.g., the papers [9, 129, 130] concerning the problems (1.1.1)–(1.1.2)). It is especially interesting to develop methods from numerical analysis for such problems, since a lot of models (for example, various physical models, models of population dynamic, etc.; see [125, 129]) can be described by abstract parabolic equations. Conditions under which the semidiscretization can be examined are very general. We introduce the following notions.

Definition 1.3. A sequence of elements $\{x_n\}$, $x_n \in E_n$, $n \in \mathbb{N}$, is said to be $\mathcal{P}$-*compact* if for any $\mathbb{N}' \subseteq \mathbb{N}$, there exist $\mathbb{N}'' \subseteq \mathbb{N}'$ and $x \in E$ such that $x_n \xrightarrow{\mathcal{P}} x$ as $n \to \infty$ in $\mathbb{N}''$.

Definition 1.4. A function $\mu(\cdot)$ is called a *discrete measure of noncompactness* if for any bounded sequence $\{x_n\}$, $x_n \in E_n$, one has

$$\mu(\{x_n\}) = \inf\Big\{\epsilon > 0 : \forall \mathbb{N}' \subseteq \mathbb{N} \ \exists \mathbb{N}'' \subseteq \mathbb{N}' \ \& \ x' \in E$$

$$\text{such that } \|x_n - p_n x'\| \leq \epsilon, \ n \in \mathbb{N}''\Big\}.$$

Definition 1.5. We say that a sequence of operators $\{\mathcal{B}_n\}$, $\mathcal{B}_n : E_n \to E_n$, *converges compactly* to an operator $\mathcal{B}$, $\mathcal{B} : E \to E$, if $\mathcal{B}_n \xrightarrow{\mathcal{PP}} \mathcal{B}$ discretely and the following compactness condition holds:

$$\|x_n\| = O(1) \quad \text{implies that} \quad \{\mathcal{B}_n x_n\} \text{ is } \mathcal{P}\text{-compact.}$$

Definition 1.6. A solution $v(\cdot)$ of the Cauchy problem (1.1.1) is said to be *stable in the Lyapunov sense* if for any $\epsilon > 0$, there is $\delta > 0$ such that the inequality

$$\big\|v(0) - \tilde{v}(0)\big\| \leq \delta$$

implies

$$\max_{0 \leq t < \infty} \big\|v(t) - \tilde{v}(t)\big\| \leq \epsilon,$$

where $\tilde{v}(\cdot)$ is a mild solution of (1.1.1) with the initial value $\tilde{v}(0)$.

Definition 1.7. A solution $v(\cdot)$ of the Cauchy problem (1.1.1) is said to be *uniformly asymptotically stable* at a point $v(0)$ if it is stable in the Lyapunov sense and, moreover, for any mild solution $\tilde{v}(\cdot)$ of (1.1.1) with $\left\| v(0) - \tilde{v}(0) \right\| \leq \delta$, we have

$$\lim_{t \to \infty} \left\| v(t) - \tilde{v}(t) \right\| = 0$$

uniformly in $\tilde{v}(\cdot) \in B(v(0); \delta)$, i.e., there exists a function $\phi_{v(0),\delta}(\cdot)$ such that

$$\left\| v(t; v(0)) - v(t; \tilde{v}(0)) \right\| \leq \phi_{v(0),\delta}(t),$$

where $\phi_{v(0),\delta}(t) \to 0$ as $t \to \infty$ and $\left\| v(0) - \tilde{v}(0) \right\| \leq \delta$.

Remark 1.4. Sufficient conditions of the existence of a uniformly asymptotically stable solution of the problem (1.1.1) are given, for example, in [129, Theorem 8.1.1]. They concern the location of the spectrum of the operator $A + \dfrac{\partial f}{\partial v}(t, v^*(t))$.

Constructive conditions on an operator A and a function F ensuring that the problem (1.2.16) is asymptotically k-dimensional are given in [249, 250]. These conditions concern the location of eigenvalues of the operator A, namely, $\lambda_{k+1} - \lambda_k > 2L$, $\lambda_{k+1} > L$. On the hyperbolicity properties of inertial manifolds, see [253]. For the development of this topic, see [251, 252, 254, 255].

Now we recall the following theorem.

Theorem 1.4 (see [57]). *Assume that Conditions* (A) *and* (B$_1$) *are fulfilled and the compact resolvents $R(\lambda; A)$ and $R(\lambda; A_n)$ converge compactly for some $\lambda \in \rho(A)$:*

$$(\lambda I_n - A_n)^{-1} \xrightarrow{\ \mathcal{PP}\ } (\lambda I - A)^{-1}.$$

Assume that the following conditions hold:

(i) *the functions f and f_n are sufficiently smooth, so that there exists an isolated mild solution $u^*(\cdot)$ of the periodic problem (1.1.2) with $u^*(0) = x^*$ such that the Cauchy problem*

$$v'(t) = Av(t) + f\big(t, v(t)\big), \quad v(0) = x^*, \tag{1.4.3}$$

has a uniformly asymptotically stable isolated solution at the point x^*;

(ii) $f_n(t, x_n) \xrightarrow{\ \mathcal{P}\ } f(t, x)$ *for any $t \in [0, T]$ as $x_n \xrightarrow{\ \mathcal{P}\ } x$.*

Then for almost all n, the problems (1.4.1) *have periodic mild solutions* $u_n^*(t)$, $t \in [0, T]$, *in a neighborhood of* $p_n u^*(\cdot)$, *where* $u^*(\cdot)$ *is a mild periodic solution of* (1.1.2) *with* $u^*(0) = x^*$. *Each sequence* $\{u_n^*(\cdot)\}$ *is* $\mathcal{P}$-*compact and* $u_n^*(t) \xrightarrow{\mathcal{P}} u^*(t)$ *uniformly with respect to* $t \in [0, T]$.

Proof. We divide the proof into several steps.

Step 1. First, we show that the compact convergence of the resolvents $R(\lambda; A_n) \xrightarrow{\mathcal{PP}} R(\lambda; A)$ is equivalent to the compact convergence of the C_0-semigroups $\exp(tA_n) \to \exp(tA)$ for any $t > 0$. Let $\|x_n\| = O(1)$. Then the estimate

$$\left\| A_n \exp(tA_n) \right\| \le \frac{M}{t} e^{\omega t}$$

implies the boundedness of the sequence $\big\{ (A_n - \lambda I_n) \exp(tA_n) x_n \big\}$. Due to the compact convergence of resolvents, we conclude the compactness of the sequence $\big\{ \exp(tA_n) x_n \big\}$.

The necessity will be proved if for a measure of noncompactness $\mu(\cdot)$ (see the definition in [235, 288, 295]), we establish that

$$\mu\Big(\big\{ (\lambda I_n - A_n)^{-1} x_n \big\} \Big) = 0 \quad \text{for } \|x_n\| = O(1).$$

We have

$$\mu\Big(\big\{ (\lambda - A_n)^{-1} x_n \big\} \Big) = \mu \left(\left\{ \int_0^\infty e^{-\lambda t} \exp(tA_n) x_n \, dt \right\} \right)$$

$$\le \mu \left(\left\{ \int_0^q e^{-\lambda t} \exp(tA_n) x_n \, dt \right\} \right)$$

$$+ \mu \left(\left\{ \int_Q^\infty e^{-\lambda t} \exp(tA_n) x_n \, dt \right\} \right)$$

$$+ \mu \left(\left\{ \exp(\epsilon A_n) \int_q^Q e^{-\lambda t} \exp\big((t - \epsilon) A_n \big) x_n dt \right\} \right).$$

One can make the first two terms smaller than ϵ by an approproate choice of q and Q. The last term is equal to zero due to the compact convergence

$$\exp(\epsilon A_n) \xrightarrow{\mathcal{PP}} \exp(\epsilon A) \quad \text{for any } 0 < \epsilon < q.$$

Step 2. The operator K defined by (1.1.5) is compact in the space F. Indeed, following [225, p. 192], we see that the operator

$$\mathcal{F}_\epsilon(u_k)(t) = \exp(\epsilon A) \int_0^{t-\epsilon} \exp\big((t - s - \epsilon) A \big) f\big(s, u_k(s) \big) ds$$

maps any bounded set of functions $\{u_k(\cdot)\}$, $\|u_k(\cdot)\|_F \leq C$, to a compact set in E for any $t > 0$ and $0 < \epsilon < t$. We see that

$$\big\|\mathcal{F}_\epsilon(u_k)(t) - \mathcal{F}(u_k)(t)\big\| \leq C\epsilon \quad \text{for any } t \in (0, T],$$

where

$$\mathcal{F}(u_k)(t) = \int_0^t \exp\big((t - s)A\big) f\big(s, u_k(s)\big) ds$$

and $0 < \epsilon < t$. Then we conclude that the operator $\mathcal{F}(\cdot)(t) : F \to E$ is compact for the same $t > 0$. For $t = 0$, the operator $\mathcal{F}(\cdot)(0)$ is also compact. Moreover, the set of functions $\{F_k(\cdot)\}$, $F_k(t) = \mathcal{F}(u_k)(t)$, $t \in [0, T]$, is an equibounded and equicontinuous family since for $0 < t_1 < t_2$, we have

$$\big\|F_k(t_2) - F_k(t_1)\big\|$$
$$\leq C \left(\int_0^{t_1} \big\| \exp\big((t_2 - s)A\big) - \exp\big((t_1 - s)A\big)\big\| ds + |t_2 - t_1| \right)$$

and $\exp(\cdot A)$ is uniformly continuous for $t > 0$.

The sequence $\{y_k\}$, where

$$y_k = \big(I - \exp(TA)\big)^{-1} \int_0^T \exp\big((T - s)A\big) f\big(s, u_k(s)\big) ds \in E,$$

is compact since $\{\mathcal{F}(u_k)(T)\}$ is a compact set. Thus, $\{\exp(\cdot A)y_k\}$ is a compact sequence of functions in F. The generalized Arzelà–Ascoli theorem implies that the operator K is compact.

Step 3. Now we consider the operators K_n defined by (1.4.2) on the spaces

$$F_n = C([0, T]; E_n) \equiv \Big\{u_n(t) : \|u_n\|_{F_n} = \max_{t \in [0, T]} \big\|u_n(t)\big\|_{E_n} < \infty\Big\}.$$

It is easy to see that $K_n \xrightarrow{\mathcal{PP}} K$. Indeed,

$$I_n \xrightarrow{\mathcal{PP}} I \text{ stably}, \quad \exp(TA_n) \xrightarrow{\mathcal{PP}} \exp(TA) \text{ compactly};$$

therefore,

$$I_n - \exp(TA_n) \to I - \exp(TA) \text{ regularly},$$

the null-space $\mathcal{N}\big(I - \exp(TA)\big)$ is trivial, i.e., $\mathcal{N}\big(I - \exp(TA)\big) = \{0\}$, and $I_n - \exp(TA_n)$ are Fredholm operators with zero index. Then it follows from [288] that

$$I_n - \exp(TA_n) \xrightarrow{\mathcal{PP}} I - \exp(TA) \text{ stably},$$

i.e.,

$$\left(I_n - \exp(TA_n)\right)^{-1} \xrightarrow{\mathcal{PP}} \left(I - \exp(TA)\right)^{-1}$$

and the convergence $K_n \xrightarrow{\mathcal{PP}} K$ follows from the majorant convergence theorem. To show that $K_n \to K$ compactly, we assume that $\|u_n\|_{F_n} = O(1)$. We conclude that $\{K_n u_n\}$ is $\mathcal{P}$-compact by the generalized Arzelà–Ascoli theorem (see [169] or [225, p. 192–193]). To prove this, we verify the fact of vanishing of the noncompactness measure $\mu\left(\{(K_n u_n)(t)\}\right) = 0$ for all $t \in [0, T]$. Consider the equality

$$(K_n u_n)(t) = \exp(tA_n)y_n + \psi_n^\tau(t) + \varphi_n^\tau(t),$$

where

$$y_n = \left(I_n - \exp(TA_n)\right)^{-1} \int_0^T \exp\left((T-s)A_n\right)f_n\left(s, u_n(s)\right)ds,$$

$$\psi_n^\tau(t) = \exp(\tau A_n) \int_0^{t-\tau} \exp\left((t-s-\tau)A_n\right)f_n\left(s, u_n(s)\right)ds,$$

$$\varphi_n^\tau(t) = \int_{t-\tau}^t \exp\left((t-s)A_n\right)f_n\left(s, u_n(s)\right)ds.$$

Due to the boundedness of $\left\|f_n(\cdot, u_n(\cdot))\right\|_{F_n}$ we can choose the term $\|\varphi_n^\tau(\cdot)\|_{F_n}$ sufficiently small with sufficiently small τ and $\mu\left(\{\psi_n^\tau\}\right) = 0$. The sequence $\{y_n\}$ is $\mathcal{P}$-compact and we are done.

Step 4. The condition of the existence of an isolated, uniformly assymptotically stable solution $v(t; x^*)$ of the problem (1.4.3) implies that in a small neighborhood of x^* (say, in $\mathcal{U}(x^*, \rho) \subset E$), the operator $\mathcal{K}$ is compact, since the set $\mathcal{F}(u_k)(T)$ is compact for any $\{u_k\}$, $u_k(t) \in \mathcal{U}(x^*, \epsilon)$, $t \in [0, T]$, with $\|u_k(0) - x^*\| \leq \delta$. The point x^* is an isolated zero of the compact vector field $I - \mathcal{K}$ and $\mathrm{ind}(x^*; I - \mathcal{K})$ is defined. Similarly, the function $u^*(t) = v(t; x^*)$, $t \in [0, T]$, which is a solution of the problem (1.1.5), is an isolated zero of the field $I - K$ and $\mathrm{ind}(u^*(\cdot); I - K)$ is defined.

Step 5. The condition of the uniform asymptotical stability of a solution $v(\cdot)$ of the problem (1.1.1) at a point x^* implies that there is an integer number m such that the operator $\mathcal{K}^m$ maps the ball $\mathcal{U}(x^*, \delta)$ into itself; more precisely,

$$\left\|\mathcal{K}^m(x^*) - \mathcal{K}^m(x)\right\| \leq \phi_{x^*, \delta}(mT) < \delta \quad \text{for any } x \in \mathcal{U}(x^*, \delta).$$

This means that $\mathrm{ind}(x^*; I - \mathcal{K}^m) = 1$ and by [168, Theorem 31.1] we obtain that $\mathrm{ind}(x^*; I - \mathcal{K}) = 1$. Using Step 4 and the affinity principle, we have

$$\mathrm{ind}(u^*(\cdot); I - K) = 1.$$

Thus, $K_n \to K$ compactly and $\mathrm{ind}(u^*(\cdot); I - K) = 1$; applying a result from [288], we conclude that the set of solutions of the problems (1.4.2) is nonempty, any sequence of solutions $\{u_n^*(\cdot)\}$ is $\mathcal{P}$-compact and, moreover, $u_n^*(t) \xrightarrow{\mathcal{PP}} u^*(t)$ uniformly with respect to $t \in [0, T]$ as $n \to \infty$. The theorem is proved. $\qquad\square$

Remark 1.5. Condition (i) of Theorem 1.4 can be changed by one of the following two conditions:

(i′) there is an isolated fixed point x^* of the operator $\mathcal{K}$ such that
$$\mathrm{ind}(x^*, I - \mathcal{K}) \neq 0$$

  or

(i″) there is an isolated fixed point $u^*(\cdot)$ of the operator K such that
$$\mathrm{ind}(u^*(\cdot), I - K) \neq 0.$$

Recall that each mild solution of the problem (1.2.16) satisfies the equation

$$u(t) = (\tilde{K}u)(t) \equiv \exp(tA)u^0 + \int_0^t \exp\left((t - s)A\right) f(u(s)) \, ds, \qquad (1.4.4)$$

where $t \in [0, T]$, $T < \infty$.

We also consider in the Banach spaces E_n^θ the following family of parabolic problems:

$$\begin{aligned} u_n'(t) &= A_n u_n(t) + f_n(u_n(t)), \quad t \in \mathbb{R}^+, \\ u_n(0) &= u_n^0 \in E_n^\theta, \end{aligned} \qquad (1.4.5)$$

where $u_n^0 \xrightarrow{\mathcal{P}} u^0$, the operators A_n and A are consistent, and the functions $f_n(\cdot) : E_n^\theta \to E_n$ are bounded and Lipschitz continuous. We also assume that $f_n(x_n) \xrightarrow{\mathcal{P}} f(x)$ and $f_n'(x_n) \xrightarrow{\mathcal{P}^\theta \mathcal{P}} f'(x)$ as $x_n \xrightarrow{\mathcal{P}^\theta} x$.

According to Theorem 1.3, assume that the operators A_n generate (see (B_1)) analytic C_0-semigroups satisfying the estimate

$$\|e^{tA_n}\| \leq M e^{-t\tilde{\omega}}, \quad t \geq 0,$$

where $\tilde{\omega} > 0$, the resolvents $(\lambda I_n - A_n)^{-1}$ are compact for some $\lambda \in \rho(A_n)$ and converge compactly for some $\lambda \in \bigcap_{n=1}^{\infty} \rho(A_n) \cap \rho(A)$:

$$(\lambda I_n - A_n)^{-1} \xrightarrow{\mathcal{PP}} (\lambda I - A)^{-1},$$

and the functions $f_n(\cdot) : E_n^\theta \to E_n$ are globally Lipschitzian and bounded, i.e., there is a constant $\tilde{c} > 0$ such that

$$\|f_n(w_n)\|_{E_n} \leq \tilde{c} \quad \text{for all } w_n \in E_n^\theta,$$

and they are continuously Fréchet differentiable functions.

Let $t \mapsto u_n(t, u_n^0)$ be a solution of (1.4.5). It is known that all solutions of (1.4.5) are globally defined; hence $t \mapsto u_n(t, \cdot)$ is a continuous, nonlinear semigroup for each $n \in \mathbb{N}$. Consider the operator equations

$$u_n(t) = (\tilde{K}_n u_n)(t) \equiv \exp(tA_n)u_n^0 + \int_0^t \exp\big((t-s)A_n\big) f_n(u_n(s))\, ds, \quad (1.4.6)$$

where $t \in [0, T]$, $T < \infty$. Consider also the operators $\tilde{K}$ and $\tilde{K}_n$ defined by (1.4.4) and (1.4.6) on the spaces

$$F = C([0, T]; E^\theta) \equiv \Big\{ u(t) : \|u\|_F = \max_{t \in [0, T]} \|u(t)\|_{E^\theta} < \infty \Big\},$$

$$F_n = C([0, T]; E_n^\theta) \equiv \Big\{ u_n(t) : \|u_n\|_{F_n} = \max_{t \in [0, T]} \|u_n(t)\|_{E_n^\theta} < \infty \Big\}.$$

Theorem 1.5. *Assume that the resolvent $(\lambda I - A)^{-1}$ is compact for some $\lambda \in \rho(A)$ and the function $f(\cdot)$ is sufficiently smooth. Then the operator $\tilde{K}$ is compact.*

Proof. The operator $\tilde{K}$ defined by (1.4.4) is compact in the space F. Indeed, we obtain that the operator

$$\mathcal{F}_\epsilon(u^k)(t) = \exp(\epsilon A) \int_0^{t-\epsilon} \exp\big((t - s - \epsilon)A\big) f\big(u^k(s)\big)\, ds$$

maps any bounded set of functions $\{u^k(\cdot)\}$, $\|u^k(\cdot)\|_F \leq C$, into a compact set in E for any $t > 0$ and $0 < \epsilon < t$. We see that

$$\|\mathcal{F}_\epsilon(u^k)(t) - \mathcal{F}(u^k)(t)\| \leq C\epsilon$$

for any $t \in (0, T]$, where

$$\mathcal{F}(u^k)(t) = \int_0^t \exp\big((t - s)A\big) f\big(u^k(s)\big)\, ds$$

and $0 < \epsilon < t$. This implies that the operator $\mathcal{F}(\cdot)(t) : F \to E^\theta$ is compact for any $t > 0$. For $t = 0$, the operator $\mathcal{F}(\cdot)(0)$ is also compact. Moreover, the set of functions $\{F_k(\cdot)\}$, $F_k(t) = \mathcal{F}(u^k)(t)$, $t \in [0, T]$, is an equibounded and equicontinuous family, since for $0 < t_1 < t_2$ we obtain

$$\big\|F_k(t_2) - F_k(t_1)\big\|_{E^\theta}$$

$$\leq C \left(\int_0^{t_1} \big\| \exp\big((t_2 - s)A\big) - \exp\big((t_1 - s)A\big) \big\|_{E^\theta}\, ds + |t_2 - t_1|^{1-\theta} \right)$$

and $\exp(\cdot A)$ is uniformly continuous in $t > 0$. By the generalized Arzelà–Ascoli theorem, similarly to Theorem 1.6, we conclude that the operator $\tilde{K}$ is compact. □

Proposition 1.1. *Let Conditions* (A) *and* (B$_1$) *be fulfilled. Then the compact convergence of the resolvents*

$$(\lambda I_n - A_n)^{-1} \xrightarrow{\mathcal{PP}} (\lambda I - A)^{-1} \qquad (1.4.7)$$

is equivalent to the compact convergence of the analytic C_0-semigroups $\exp(tA_n) \xrightarrow{\mathcal{PP}} \exp(tA)$ *for any $t > 0$.*

Proof. First, we show that the compact convergence of the resolvents (1.4.7) is equivalent to the compact convergence of the C_0-semigroups $\exp(tA_n) \to \exp(tA)$ for any $t > 0$. Let $\|x_n\| = O(1)$. Then from the estimate

$$\left\| A_n \exp(tA_n) \right\| \leq \frac{M}{t} e^{\omega t}$$

we obtain the boundedness of the sequence $\left\{ (A_n - \lambda I_n) \exp(tA_n) x_n \right\}$. Due to Theorem 2.2 and the compact convergence of the resolvents, we obtain the compactness of the sequence $\left\{ \exp(tA_n) x_n \right\}$.

The necessity will be proved if for a noncompactness measure $\mu(\cdot)$ we establish that

$$\mu\left(\left\{ (\lambda I_n - A_n)^{-1} x_n \right\} \right) = 0 \quad \text{for } \|x_n\| = O(1).$$

We have

$$\mu\left(\left\{ (\lambda I_n - A_n)^{-1} x_n \right\} \right)$$
$$= \mu\left(\left\{ \int_0^\infty e^{-\lambda t} \exp(tA_n) x_n \, dt \right\} \right)$$
$$\leq \mu\left(\left\{ \int_0^q e^{-\lambda t} \exp(tA_n) x_n \, dt \right\} \right)$$
$$+ \mu\left(\left\{ \int_Q^\infty e^{-\lambda t} \exp(tA_n) x_n \, dt \right\} \right)$$
$$+ \mu\left(\left\{ \exp(\epsilon A_n) \int_q^Q e^{-\lambda t} \exp\left((t - \epsilon) A_n\right) x_n \, dt \right\} \right).$$

The first two terms of the right-hand side can be made less than ϵ by an appropriate choice of q and Q. The last term is equal to zero due to the compact convergence $\exp(\epsilon A_n) \xrightarrow{\mathcal{PP}} \exp(\epsilon A)$ for any $0 < \epsilon < q$. □

Theorem 1.6. *The operator convergence* $\tilde{K}_n \to \tilde{K}$ *is compact.*

Proof. To show that $\tilde{K}_n \to \tilde{K}$ compactly, we assume that $\|u_n\|_{F_n} = O(1)$. Now $\{\tilde{K}_n u_n\}$ is $\mathcal{P}$-compact by the generalized Arzelà–Ascoli theorem. To show this, we prove the vanishing of the noncompactness measure $\mu\big(\{(\tilde{K}_n u_n)(t)\}\big) = 0$ for all $t \in [0, T]$. Consider the relation

$$(\tilde{K}_n u_n)(t) = \exp(tA_n)u_n^0 + \psi_n^\tau(t) + \varphi_n^\tau(t),$$

where

$$\psi_n^\tau(t) = \exp(\tau A_n) \int_0^{t-\tau} \exp\big((t - s - \tau)A_n\big) f_n\big(u_n(s)\big) ds,$$

$$\varphi_n^\tau(t) = \int_{t-\tau}^t \exp\big((t - s)A_n\big) f_n\big(u_n(s)\big) ds.$$

Due to the boundedness of $\big\|f_n\big(u_n(\cdot)\big)\big\|_{F_n}$, we can choose the term $\|\varphi_n^\tau(\cdot)\|_{F_n}$ sufficiently small with sufficiently small τ and $\mu(\{\psi_n^\tau\}) = 0$. The sequence $\{e^{tA_n}u_n^0\}$ is also $\mathcal{P}$-compact.

Clearly, $\{\tilde{K}_n u_n\}$ is uniformly bounded in n. Prove the equicontinuity; for $t_1 \le t_2$, we have

$$\left\| A_n^\theta \int_0^{t_2} \exp\big((t_2 - s)A_n\big) f_n\big(u_n(s)\big) ds \right.$$

$$\left. - A_n^\theta \int_0^{t_1} \exp\big((t_1 - s)A_n\big) f_n\big(u_n(s)\big) ds \right\|_{E_n}$$

$$\le \left\| \int_{t_1}^{t_2} A_n^\theta e^{(t_2 - s)A_n} f_n\big(u_n(s)\big) ds \right\|_{E_n}$$

$$+ \left\| \int_0^{t_1} A_n^\theta \big(e^{(t_2 - s)A_n} - e^{(t_1 - s)A_n}\big) f_n\big(u_n(s)\big) ds \right\|_{E_n}.$$

The first term tends to zero as $|t_2 - t_1| \to 0$ since

$$\int_{t_1}^{t_2} \frac{ds}{(t_2 - s)^\theta} = -\frac{(t_2 - s)^{1-\theta}}{1 - \theta} \Big|_{s=t_1}^{s=t_2} = \frac{(t_2 - t_1)^{1-\theta}}{1 - \theta}.$$

To estimate the second term for any $\epsilon > 0$ and some $0 < \delta \le \epsilon^{1/(1-\theta)}$, we consider the cases $t_1 \le \delta$ and $t_1 > \delta$ separately. If $t_1 \le \delta$ and $|t_2 - t_1| \le \delta$, then

$$\left\| \int_0^{t_2} \cdots ds \right\| \le c\delta^{1-\theta} \le C\epsilon.$$

If $t_1 > \delta$, then one can partition the second term as follows:

$$\int_0^{t_1} = \int_0^{t_1-\delta} + \int_{t_1-\delta}^{t_1}.$$

So, if $|t_2 - t_1| \leq \delta$, then

$$\int_{t_1-\delta}^{t_1} \leq c\delta^{1-\theta} \leq C\epsilon$$

as before. Since $t_1 - s$, $t_2 - s \in [\delta, t_1]$ for $s \in [0, t_1 - \delta]$, we see that

$$\int_0^{t_1} A_n^\theta \left(e^{(t_2-s)A_n} - e^{(t_1-s)A_n} \right) f_n\big(u_n(s)\big) ds \to 0$$

as $|t_2 - t_1| \to 0$ due to the convergence

$$\left\| A_n^\theta \left(e^{(t_2-s)A_n} - e^{(t_1-s)A_n} \right) \right\|_{E_n} \to 0$$

as $|t_2 - t_1| \to 0$, i.e., the arguments $t_2 - s$ and $t_1 - s$ are separated from zero be a distance more than δ. $\qquad\square$

Theorem 1.7 (see [235]). *Assume that Conditions* (A) *and* (B$_1$) *are fulfilled and the compact resolvents* $R(\lambda; A)$ *and* $R(\lambda; A_n)$ *converge compactly for some* $\lambda \in \rho(A)$:

$$(\lambda I_n - A_n)^{-1} \xrightarrow{\mathcal{PP}} (\lambda I - A)^{-1}.$$

We also assume that Conditions (A) *and* (B$_1$), *the smoothness conditions for* f_n *and* f, *and the condition*

$$f_n(x_n) \xrightarrow{\mathcal{P}} f(x) \quad as \ x_n \xrightarrow{\mathcal{P}^\theta} x$$

are satisfied. Then for solutions of the problems (1.4.4) *and* (1.4.6) *we have*

$$u_n(t, u_n^0) \xrightarrow{\mathcal{P}^\theta} u(t, u^0) \quad uniformly \ in \ t \in [0, T], \ T < \infty, \ as \ u_n^0 \xrightarrow{\mathcal{P}^\theta} u^0.$$

Proof. We know (see Theorem 1.2) that $\mathrm{ind}(u^*, I - \tilde{K}) = 1$. By Theorem 1.6, we have the compact operator convergence $\tilde{K}_n \to \tilde{K}$, therefore, according to [288, Theorem 3] we have

$$\gamma(I_n - \tilde{K}_n, \partial\Omega_n) = 1 \quad \text{for } n \geq n_0.$$

This means that there is a sequence $\{u_n^*\}$ such that, for each $n \in \mathbb{N}$, u_n^* is a solution of the problems (1.4.6) and $\{u_n^*\}$, $u_n^* \in F_n$, is compact. Hence we conclude that $u_n^*(t) \xrightarrow{\mathcal{P}^\theta} u^*(t)$ uniformly in $t \in [0, T]$ as $n \to \infty$. $\qquad\square$

Chapter 2

Convergence of Attractors

The interest in studying the existence and approximation of attractors continues unabated (see [6, 39, 55, 80, 83, 133, 262, 298, 317, 318]). A standard result in the theory of dynamical systems says that the attractors of parametric dynamical systems converge upper semicontinuously in the parameter (see [68, 125]). Under additional assumptions, the lower semicontinuous covergence and hence the continuous convergence of the attractors holds. Such assumptions typically involve the structure of the attractors, e.g., the Morse–Smale structure. There is a way (see [183]) to state that the continuous convergence of attractors is equivalent to their property to be equiattracting (see also [159], where a similar result was presented without proof).

The same results on the upper semicontinuous convergence hold for numerical approximations of ordinary differential equations (see [24, 127, 153, 270]) and also for Galerkin approximations of parabolic partial differential equations (see [154]). The convergence of attractors of nonconformal finite-element approximations of parabolic partial differential equations and the dependence of attractors of delay differential equations on the delay are more complicated since systems being compared have different state spaces, but a form of analogous upper semicontinuous convergence can be established (see [152, 204]).

In this chapter, we use the concept of "discrete convergence" defined in Chapter 1 for comparing systems on different spaces to prove the equivalence of equiattraction and continuous convergence of attractors of such systems. This result will be applied to general approximation schemes for abstract semilinear parabolic systems, which include finite-difference and nonconformal finite-element approximations as well as more straightforward Galerkin and conformal finite-element approximations.

2.1   Discrete Convergence of Sets

For nonempty subsets U_n and V_n of E_n, $n \in \mathbb{N}$, and a nonempty subset U of E, we introduce the Hausdorff semi-distance

$$d_n(U_n, V_n) = \sup_{u_n \in U_n} \inf_{v_n \in V_n} \left\| u_n - v_n \right\|_{E_n}$$

and the Hausdorff distance

$$\delta_n(U_n, U) \equiv \delta_n(U, U_n) = \max \left\{ d_n(p_n U, U_n),\ d_n(U_n, p_n U) \right\}.$$

Note that $\delta_n(U_n, U)$ can be equal to zero without U_n and U being equal.

Definition 2.1. Let $\mathcal{A}_n \subset E_n$ for $n \in \mathbb{N}$ and $\mathcal{A} \subset E$.

(1) We say that a family of sets $\{\{\mathcal{A}_n\},\ n \in \mathbb{N}\}$ *converges $\mathcal{P}$-upper semi-continuously* to a set $\mathcal{A}$ in E if $d_n(\mathcal{A}_n, p_n\mathcal{A}) \to 0$ as $n \to \infty$.
(2) We say that a family of sets $\{\{\mathcal{A}_n\},\ n \in \mathbb{N}\}$ *converges $\mathcal{P}$-lower semi-continuously* to a set $\mathcal{A}$ in E if $d_n(p_n\mathcal{A}, \mathcal{A}_n) \to 0$ as $n \to \infty$.

Combining these notions, we say that $\mathcal{A}_n$ *converges $\mathcal{P}$-continuously* to $\mathcal{A} \subset E$ as $n \to \infty$ if

$$\delta_n(\mathcal{A}_n, p_n\mathcal{A}) \to 0 \quad \text{as } n \to \infty. \tag{2.1.1}$$

Lemma 2.1 (see [69]). *Let $\mathcal{A}_n \subset E_n$ for $n \in \mathbb{N}$ and $\mathcal{A} \subset E$.*

(1) *If any sequence $\{u_n\}$ with $u_n \in \mathcal{A}_n$ possesses a $\mathcal{P}$-convergent subsequence with limit belonging to $\mathcal{A}$, then $\mathcal{A}_n$ converges $\mathcal{P}$-upper semicontinuously to $\mathcal{A}$.*
(2) *If $\mathcal{A}$ is compact and for any $u \in \mathcal{A}$, there is a sequence $\{u_n\}$ with $u_n \in \mathcal{A}_n$, which $\mathcal{P}$-converges to u, then $\mathcal{A}_n$ converges $\mathcal{P}$-lower semicontinuously to $\mathcal{A}$.*

2.2   Equiattraction and Continuous Convergence

A semidynamical system $S_n(\cdot)$ on a Banach space $(E_n, \| \cdot \|_{E_n})$ is said to be *dissipative* (see [155]) if there is a bounded subset U_n of E_n such that for every bounded subsets B_n of E_n, there exists a finite time $T_n = T_n(B_n)$ such that

$$S_n(t)B_n \subseteq U_n \quad \text{for } t \geq T_n.$$

Definition 2.2. We say that a family of semidynamical systems $\{S_n(\cdot)\}$, $n \in \mathbb{N}$, is *$\mathcal{P}$-equidissipative* if there is a uniformly bounded sequence $\{U_n\}$,

$n \in \mathbb{N}$, of subsets U_n of E_n such that for any uniformly bounded sequence $\mathcal{B} = \{\{B_n\}, \ n \in \mathbb{N}\}$, of subsets $B_n \subset E_n$, there exists finite $T_0 = T_0(\mathcal{B})$ independent of n such that

$$S_n(t)B_n \subseteq U_n \quad \text{for } t \geq T_0.$$

If a semidynamical system $S_n(\cdot)$ has a global attractor $\mathcal{A}_n$ for each $n \in \mathbb{N}$, then $\mathcal{A}_n$ is invariant, i.e., $S_n(t)\mathcal{A}_n = \mathcal{A}_n$ for all $t \geq 0$, and $\mathcal{A}_n$ attracts arbitrary bounded subsets of E_n under $S_n(\cdot)$, i.e., for every bounded subset B_n in E_n and every $\epsilon > 0$, there exists $\tau_n = \tau_n(B_n, \varepsilon) > 0$ such that

$$d_n(S_n(t)B_n, \mathcal{A}_n) < \varepsilon \quad \forall t \geq \tau_n. \tag{2.2.1}$$

Definition 2.3. A family of attractors $\{\mathcal{A}_n\}$, $n \in \mathbb{N}$, of a family of semidynamical systems $\{S_n(\cdot)\}$, $n \in \mathbb{N}$, is said to be $\mathcal{P}$-*equiattracting* if for each uniformly bounded sequence $\mathcal{B} = \{B_n, \ n \in \mathbb{N}\}$ of subsets B_n of E_n and every $\epsilon > 0$, there exists finite $T = T(\mathcal{B}, \epsilon) > 0$ independent of n such that

$$d_n(S_n(t)B_n, \mathcal{A}_n) < \epsilon \quad \text{for any } t \geq T. \tag{2.2.2}$$

We will assume the following convergence condition for the family $\{S_n(\cdot)\}$, $n \in \mathbb{N}$, of semidynamical systems on Banach spaces E_n and semidynamical systems $S(\cdot)$ on Banach spaces E.

Assumption 2.1. The family of semidynamical systems $\{S_n(\cdot)\}$, $n \in \mathbb{N}$, discretely compactly converges to a semidynamical system $S(\cdot)$, i.e., $S_n(t) \xrightarrow{\mathcal{PP}} S(t)$ for any $t \geq 0$ and any uniformly bounded sequence $\{x_n\}$, $n \in \mathbb{N}$, and the sequence $\{S_n(t)x_n\}$, $n \in \mathbb{N}$, is $\mathcal{P}$-compact for each $t > 0$.

The main result of this chapter is the following assertion.

Theorem 2.1 (see [155]). *Assume that a family of semidynamical systems $\{S_n(\cdot)\}$, $n \in \mathbb{N}$, on Banach spaces E_n is $\mathcal{P}$-equidissipative. In addition, assume that $S(\cdot)$ is a semidynamical system on a Banach space E such that Assumption 2.1 holds for $S_n(\cdot)$ and $S(\cdot)$. Let $\mathcal{A}_n$ be a global attractor of $S_n(\cdot)$ and $\mathcal{A}$ be a global attractor of $S(\cdot)$. The family of attractors $\{\mathcal{A}_n\}$, $n \in \mathbb{N}$, is $\mathcal{P}$-equiattracting if and only if $\mathcal{A}_n$ converge $\mathcal{P}$-continuously to the global attractor $\mathcal{A}$.*

Proof. The assumptions on the dissipativity and compactness ensure the existence of the global attractors and, moreover, the fact that the sequence of attractors $\mathcal{A}_n$ is uniformly bounded.

First, we show their $\mathcal{P}$-upper semicontinuous convergence. Assume that $x_n^* \xrightarrow{\mathcal{P}} x^*$ and $x_n^* \in \mathcal{A}_n$. Since $S_n(t)\mathcal{A}_n = \mathcal{A}_n$ for any $t > 0$, there is a sequence $\{x_n\}$, $x_n \in \mathcal{A}_n$, such that $S_n(t)x_n = x_n^*$ for any $t > 0$. The set $\{x_n\}$ is bounded due to the $\mathcal{P}$-equidissipativity; moreover, $\{x_n\}$ is $\mathcal{P}$-compact. Hence there exists $\mathbb{N}' \subset \mathbb{N}$ such that $x_n \overset{\mathcal{P}}{\longrightarrow} x$ as $n \to \infty$ in $\mathbb{N}'$ for some $x \in E$. By uniqueness of the limit, it is clear that $S_n(T)x_n \overset{\mathcal{P}}{\longrightarrow} S(T)x = x^*$ as $n \to \infty$ in $\mathbb{N}'$. The uniform boundedness of $S_n(T)x_n$ implies that for any $\epsilon > 0$, there is $T(\epsilon) > 0$ such that $S(T)x$ belongs to some bounded ball and

$$d_E(S(T)x, \mathcal{A}) < \epsilon \quad \text{for all } T > T(\epsilon).$$

Since $S(T)x = x^*$, we can choose x^* all the more close to $\mathcal{A}$ by taking T sufficiently large. Since

$$d_n(x_n^*, p_n\mathcal{A}) \le d_n\big(S_n(T)x_n, p_nS(T)x\big) + d_n\big(p_nS(T)x, p_n\mathcal{A}\big),$$

we conclude that $d_n(x_n^*, p_n\mathcal{A}) \to 0$ as $n \to \infty$ in $\mathbb{N}'$. Hence $\mathcal{A}_n$ converges $\mathcal{P}$-upper semicontinuously to $\mathcal{A}$.

Now, let a family of attractors $\{\mathcal{A}_n\}$, $n \in \mathbb{N}$, be $\mathcal{P}$-equiattracting and assume that $\mathcal{A}_n$ does not converge $\mathcal{P}$-lower semicontinuously to $\mathcal{A}$. Then there exists a sequence $x^n \in \mathcal{A}$ such that

$$d_n(p_nx^n, \mathcal{A}_n) \ge 1 > 0, \quad n \in \mathbb{N}. \tag{2.2.3}$$

Since $\mathcal{A}$ is compact, there is $\mathbb{N}' \subset \mathbb{N}$ such that $x^n \to x^*$ as $n \to \infty$ in $\mathbb{N}'$. Therefore,

$$d_n(p_nx^n, \mathcal{A}_n) \le d_n(p_nx^*, \mathcal{A}_n) + \frac{\epsilon}{3}$$

for $n \ge N(\epsilon)$ in $\mathbb{N}'$. By the $S(T)$-invariance of $\mathcal{A}$, there is an element $x \in \mathcal{A}$ such that $S(T)x = x^*$.

There is a sequence $x_n \in E_n$ with $x_n \overset{\mathcal{P}}{\longrightarrow} x$ and hence

$$S_n(T)x_n \overset{\mathcal{P}}{\longrightarrow} S(T)x \quad \text{as } n \to \infty$$

by Assumption 2.1. Since $\mathcal{A}$ is bounded, the sequence $\{x_n\}$ is also contained in a sequence of uniformly bounded balls $\mathcal{B} = \big\{\{B_n\}, \ n \in \mathbb{N}\big\}$, i.e., with $x_n \in B_n$ for each n. By the $\mathcal{P}$-equiattraction, for this uniformly bounded sequence $\mathcal{B}$, there exists $T = T(\mathcal{B}, \epsilon/3)$ such that $d_n(S_n(T)B_n, \mathcal{A}_n) \le \epsilon/3$. Finally, with $T = T(\mathcal{B}, \epsilon/3)$, we obtain

$$d_n(p_nx^n, \mathcal{A}_n) \le d_n(p_nx^*, \mathcal{A}_n) + \frac{\epsilon}{3}$$

$$\le \big\|p_nS(T)x - S_n(T)x_n\big\|_{E_n} + d_n\big(S_n(T)x_n, \mathcal{A}_n\big) + \frac{\epsilon}{3} \le \epsilon$$

for sufficiently large $n \in \mathbb{N}'$ depending on ϵ. Taking $\epsilon < 1$, we obtain a contradiction to (2.2.3). Thus, the $\mathcal{P}$-lower semicontinuous convergence of attractors holds and hence their $\mathcal{P}$-continuous convergence follows.

Now we assume that $\mathcal{A}_n$ converges $\mathcal{P}$-continuously to $\mathcal{A}$, i.e., $\delta_n(\mathcal{A}_n, \mathcal{A}) \to 0$ as $n \to \infty$. We prove that they are $\mathcal{P}$-equiattracting using the proof by contradiction. Assume that for a certain uniformly bounded sequence of sets $\{B_n\}$, $n \in \mathbb{N}$, we have

$$d_n(S_n(T)B_n, \mathcal{A}_n) \geq 1 > 0 \qquad (2.2.4)$$

for any $T > 0$ uniformly with respect to $n \in \mathbb{N}'$. Then there is a sequence $\{u_n\}$, $u_n \in B_n$, such that $d_n(S_n(T)u_n, \mathcal{A}_n) \geq 1 > 0$ uniformly in $n \in \mathbb{N}'$ for any $T > 0$. Clearly, $\{S_n(t_0)u_n\}$ is $\mathcal{P}$-compact for some $t_0 > 0$ and, moreover, there exists $u^* \in E$ such that for some infinite subset $\mathbb{N}'' \subset \mathbb{N}'$, the following $\mathcal{P}$-convergence is valid: $S_n(t_0)u_n \xrightarrow{\mathcal{P}} u^*$ as $n \in \mathbb{N}''$ tends to ∞. Hence $S_n(T)u_n \xrightarrow{\mathcal{P}} S(T - t_0)u^*$ on some subsequence $n \in \mathbb{N}''$, where $\mathbb{N}'' \subset \mathbb{N}'$. Now $d(S(T - t_0)u^*, \mathcal{A})$ can become arbitrarily small as $T - t_0$ becomes arbitrarily large. Therefore,

$$d_n(S_n(T)u_n, \mathcal{A}_n) \leq d_n\big(S_n(T)u_n, p_n S(T - t_0)u^*\big)$$
$$+ d_n\big(p_n S(T - t_0)u^*, p_n \mathcal{A}\big) + d_n(p_n \mathcal{A}, \mathcal{A}_n)$$

can be made smaller than 1 for $n \in \mathbb{N}''$ and sufficiently large T; this contradicts the inequality (2.2.4). It follows that the family of attractors $\{\mathcal{A}_n\}$, $n \in \mathbb{N}$, is $\mathcal{P}$-equiattracting. This completes the proof of Theorem 2.1. $\qquad\square$

2.3 Abstract Parabolic Problems

In this section, we consider applications of Theorem 2.1 to abstract semilinear parabolic problems.

Assume that $\theta \in (0, 1)$, an operator A satisfies the condition (1.1.3), and $f(\cdot) : E^\theta \to E$ is a globally Lipschitzian, bounded, and continuously Fréchet differentiable function. Then the semilinear, autonomous, parabolic, initial-value problem (1.2.16) has a unique mild solution defined for all $t \geq 0$ and the nonlinear solution operators $u(\cdot) = S(\cdot)u^0 : \overline{\mathbb{R}}^+ \to E^\theta$ form a strongly continuous semigroup of operators on E^θ, i.e., a semidynamical system (see [129, 295]). Moreover, the semigroup $S(\cdot)$ is given by the formula of variation of constants:

$$S(t)u^0 = \exp\big((t - t_0)A\big)S(t_0)u^0 + \int_{t_0}^t \exp\big((t - s)A\big)f(S(s)u^0)ds \quad (2.3.1)$$

(here $t \geq t_0$) and is equidissipative and compact. Thus, it has a global attractor $\mathcal{A}$ in E^θ (see [69, 72]).

We consider closed linear operators $A_n : D(A_n) \subseteq E_n \to E_n$, $n \in \mathbb{N}$, whose resolvents satisfy the inequality

$$\left\|(\lambda I_n - A_n)^{-1}\right\|_{B(E_n)} \leq \frac{M_2}{|\lambda - \omega_2|}, \quad \mathrm{Re}\,\lambda > \omega_2, \quad n \in \mathbb{N},$$

for some constants $M_2 \geq 1$ and ω_2 (see Condition (B$_1$) in Theorem 1.3). Recall (see [235]) that the condition $\Delta_{cc} \neq \varnothing$ (for the definition of the sets Δ_s, Δ_c, and Δ_{cc}, see Sec. 3.2.2) implies that the estimate (B$_1$) holds in the form

$$\left\|(\lambda I_n - A_n)^{-1}\right\|_{B(E_n)} \leq \frac{M}{1 + |\lambda|}, \quad \mathrm{Re}\,\lambda > 0, \quad n \in \mathbb{N}.$$

Therefore, the operators $(-A_n)^\theta$ are well defined. Later we also will need the operators $p_n^\theta = (-A_n)^{-\theta} p_n (-A)^\theta \in B(E^\theta, E_n^\theta)$ with the same property (1.3.1), but for the spaces E^θ and E_n^θ. The operators A_n and A are assumed to be related as described in Conditions (1.1.3), (A), and (B$_1$). The convergence $y_n \xrightarrow{\mathcal{P}^\theta} y$ is defined with respect to the spaces E_n^θ and E^θ.

Definition 2.4. We say that a sequence of closed linear operators $\{A_n\}$, $A_n \in \mathcal{C}(E_n)$, $n \in \mathbb{N}$, is *compatible* with a closed linear operator $A \in \mathcal{C}(E)$ if for each $x \in D(A)$ there is a sequence $\{x_n\}$, $x_n \in D(A_n) \subseteq E_n$, $n \in \mathbb{N}$, such that $x_n \xrightarrow{\mathcal{P}} x$ and $A_n x_n \xrightarrow{\mathcal{P}} Ax$. Briefly, we say that (A_n, A) are compatible.

We assume that the operators (A_n, A) are compatible and the functions $f_n(\cdot) : E_n^\theta \to E_n$ are globally bounded, uniformly globally Lipschitz continuous, and continuously Fréchet differentiable for all $n \in \mathbb{N}$. In addition, we assume that $f_n(x_n) \xrightarrow{\mathcal{P}} f(x)$ and $f_n'(x_n) \xrightarrow{\mathcal{P}^\theta \mathcal{P}} f'(x)$ as $x_n \xrightarrow{\mathcal{P}^\theta} x$ and for any $\rho > 0$, there exists $\delta > 0$ such that

$$\sup_{n \in \mathbb{N}} \sup_{\|w_n\|_{E_n^\theta} \leq \delta} \left\| f_n'(w_n + p_n^\theta u^*) - f_n'(p_n^\theta u^*) \right\|_{B(E_n^\theta, E_n)} \leq \rho$$

for any u^*, where $Au^* + f(u^*) = 0$.

As approximations to the initial-value problem (1.2.16), we consider the autonomous initial-value problems (1.4.5), where $u_n^0 \xrightarrow{\mathcal{P}^\theta} u^0$. Under the above assumptions, each of the problems (1.4.5) has a unique mild solution

$$u_n(\cdot) = S_n(\cdot) u_n^0 : \overline{\mathbb{R}}^+ \to E_n^\theta,$$

which generates a strongly continuous semigroup $S_n(\cdot)$ given by the formula of variation of constants:

$$S_n(t)u_n^0 = \exp\big((t-t_0)A_n\big)S_n(t_0)u_n^0 + \int_{t_0}^t \exp\big((t-s)A_n\big)f_n(S_n(s)u_n^0)ds$$

$$(2.3.2)$$

(here $t \geq t_0$) for each $n \in \mathbb{N}$. Moreover, the corresponding semidynamical system $S_n(\cdot)$ is dissipative and compact and has a global attractor $\mathcal{A}_n$ in E_n^θ.

We consider some cases in which Theorem 2.1 on the equivalence of $\mathcal{P}$-equiattraction and $\mathcal{P}$-continuous convergence can be applied to the above systems in the spaces E_n^θ and E^θ. For this, it is convenient to restrict the approximation operators A_n to operators with compact resolvents.

For the definition of the sets Δ_s, Δ_c, and Δ_{cc}, see Sec. 3.2.2.

Theorem 2.2 (see [235]). *Assume that $\Delta_{cc} \neq \varnothing$. Then for any $\zeta \in \Delta_s$, the following implication holds:*

$$\|x_n\|_{E_n} = O(1) \ \& \ \big\|(\zeta I - A_n)x_n\big\|_{E_n} = O(1) \implies \{x_n\} \text{ is } \mathcal{P}\text{-compact.}$$

$$(2.3.3)$$

Conversely, if the implication (2.3.3) holds for some $\zeta \in \Delta_c \cap \rho(A)$, then $\Delta_{cc} \neq \varnothing$.

Proof. Let $(\mu I_n - A_n)^{-1} \xrightarrow{\ \mathcal{PP}\ } (\mu I - A)^{-1}$ compactly for some $\mu \in \Delta_{cc}$. Then for $\|x_n\|_{E_n} = O(1)$ and $\|(\zeta I - A_n)x_n\|_{E_n} = O(1)$, from the Hilbert identity

$$(\zeta I - A_n)^{-1} - (\mu I - A_n)^{-1} = (\mu - \zeta)(\zeta I - A_n)^{-1}(\mu I - A_n)^{-1}, \quad (2.3.4)$$

we obtain

$$x_n = (\mu I - A_n)^{-1}(\zeta I - A_n)x_n - (\zeta - \mu)(\mu I - A_n)^{-1}x_n$$

and hence $\{x_n\}$ is $\mathcal{P}$-compact.

Conversely, let the implication (2.3.3) hold for some $\zeta_0 \in \Delta_c \cap \rho(A)$. We show that $\zeta_0 \in \Delta_{cc}$. Taking a bounded sequence $\{y_n\}$, $n \in \mathbb{N}$, we obtain

$$\big\|(\zeta_0 I - A_n)^{-1}y_n\big\|_{E_n} = O(1), \quad n \in \mathbb{N}.$$

Let us apply the implication (2.3.3) to the sequence $x_n = (\zeta_0 I - A_n)^{-1}y_n$. It is easy to see that $\{x_n\}$ is $\mathcal{P}$-compact. Hence $\zeta_0 \in \Delta_{cc}$. □

We make the following assumption:

Assumption 2.2. Assume that the resolvent conditions (1.1.3) and (B$_1$) hold, $\Delta_{cc} \neq \varnothing$, the resolvents $(\lambda I_n - A_n)^{-1}$ are compact for $\lambda \in \Delta_{cc}$, and $f_n(x_n) \xrightarrow{\ \mathcal{P}\ } f(x)$ as $x_n \xrightarrow{\ \mathcal{P}^\theta\ } x$.

Theorem 2.3. *Assume that Condition 2.2 is fulfilled. Then*

$$S_n(t)u_n^0 \xrightarrow{\ \mathcal{P}^\theta\ } S(t)u^0 \quad \text{uniformly in } t \in [0,T] \text{ as } u_n^0 \xrightarrow{\ \mathcal{P}^\theta\ } u^0,$$

where $T < \infty$.

One can see that under Assumption 2.2, the family of operators $S_n(t)$ converges compactly to $S(t)$ for any $t > 0$, so the discrete compactness condition in Assumption 2.1 is satisfied.

The reader is referred to [69] (see also the references therein), where specific approximations involving finite-element and finite-difference methods are discussed in detail. In particular, it was shown how to use (2.3.3) for the proof of the compact convergence of resolvents. One can find there also a technique for checking the assumption $\Delta_{cc} \neq \varnothing$, for example, in the case of the finite-element method.

Remark on the continuous convergence and equiattraction

The continuous convergence of attractors is known to hold in certain problems if the original attractor is a Morse–Smale attractor, i.e., if it consists of a finite number of hyperbolic equilibria and their unstable manifolds, although continuous convergence is also possible for nonhyperbolic equilibria (see [159, 166]).

Here we consider a special situation. We assume that the trajectory $S(\cdot)u^0$ of the original system starting from any point u^0 converges to some limit, i.e., $S(t)u^0 \to u^*$ as $t \to \infty$. Then u^* (which may depend on u^0) is a solution of the equation $Au + f(u) = 0$, i.e., it is an equilibrium point. Such equilibria exist under the assumptions on the function $f(\cdot)$, and the problems (1.4.5) also have equilibrium points that are closed to the singleton set $\{u^*\}$, i.e., with $u_n^* \xrightarrow{\ \mathcal{P}\ } u^*$ as $n \to \infty$. In a sense, we have a form of continuity of attractors at the equilibrium points $\{u^*\}$. Assume that $a \in \mathcal{A}$ is not an equilibrium point. Then there exists a point a' in an ϵ-neighbourhood $\mathcal{U}(u^*, \epsilon)$ of u^* such that $S(T)a' = a$ for some $T > 0$. Following [69], we can find points $a_n' \in \mathcal{A}_n \cap \mathcal{U}(u_n^*, \epsilon)$ such that $a_n' \xrightarrow{\ \mathcal{P}^\theta\ } a'$ and, by the convergence assumption (Assumption 2.1), we have $a_n = S_n(T)a_n' \xrightarrow{\ \mathcal{P}^\theta\ } S(T)a' = a$. Since $a_n' \in \mathcal{A}_n$, it follows that $a_n \in \mathcal{A}_n$, and the continuity of attractors is proved. By Theorem 2.1, we can then conclude that nonlinear semigroups $S_n(\cdot)$ are $\mathcal{P}$-equiattracting.

Chapter 3

Convergence of Attractors in the Hyperbolic Case

Let E be a complex Banach space and $A : D(A) \subset E \to E$ be a closed linear operator with compact resolvent such that the condition (1.1.3) holds. In this situation,

$$\operatorname{Re} \sigma(A) = \Big\{ \operatorname{Re} \lambda : \lambda \in \sigma(A) \Big\} \subset (-\infty, 0).$$

It is well known (see [129]) that under these assumptions, there are positive constants M and β such that

$$\|e^{tA}\|_{B(E^{\theta_1}, E^{\theta_2})} \le M t^{\theta_1 - \theta_2} e^{-\beta t}, \quad \theta_2 > \theta_1 \ge 0, \quad t \ge 0. \tag{3.0.1}$$

Under the above assumptions and the assumption that $f(\cdot) : E^{\theta} \to E$ is a globally bounded, uniformly globally Lipschitz continuous, and continuously Fréchet differentiable function, a mild solution of the problem (1.2.16) is globally defined (see [129]). Let $u(t) = S(\cdot)u^0 : \overline{\mathbb{R}}^+ \to E^{\theta}$ be a mild solution of the problem (1.2.16). Then we can define a family of nonlinear operators $\{S(t) : t \ge 0\}$ possessing the following properties:

(i) $S(0) = I$,
(ii) $S(t + s) = S(t)S(s)$, $t \ge 0$, $s \ge 0$,
(iii) $\overline{\mathbb{R}}^+ \times E^{\theta} \ni (t, x) \mapsto S(t)x \in E^{\theta}$ is continuous.

It is well known that the nonlinear semigroup $S(\cdot)$ is given by the formula of variation of constants (2.3.1). For each $u^0 \in E^{\theta}$, the positive orbit $\gamma^+(u^0)$ passing through u^0 is defined as $\gamma^+(u^0) = \{S(t)u^0 : t \ge 0\}$. A negative orbit passing through u^0 is a continuous function $\phi : (-\infty, 0] \to E^{\theta}$ such that $\phi(0) = u^0$ and, for any $s \le 0$, $S(t)\phi(s) = \phi(t + s)$ for $0 \le t \le -s$. A complete orbit passing through u^0 is a function $\phi : \mathbb{R} \to E^{\theta}$ such that $\phi(0) = u^0$ and, for any $s \in \mathbb{R}$, $S(t)\phi(s) = \phi(t + s)$ for $t \ge 0$.

3.1   Attractors for Abstract Parabolic Problems

We assume that an operator A generates (see (1.1.3)) an analytic C_0-semigroup satisfying the estimate

$$\|e^{tA}\| \le M e^{-t\omega}, \quad t \ge 0,$$

where $\omega > 0$, the resolvent $(\lambda I - A)^{-1}$ is compact for some $\lambda \in \rho(A)$, and $f(\cdot) : E^\theta \to E$ is a globally Lipschitzian, bounded (i.e., there is a constant $K > 0$ such that $\|f(x)\|_E \le K$ for all $x \in E^\theta$), and continuously Fréchet differentiable function.

3.1.1   *Equilibrium points*

Now we describe the structure of attractors for the problem (1.2.16). We start from simplest elements of attractors, namely, equilibrium solutions. Equilibrium solutions of the problem (1.2.16) are solutions independent of time: $u(t) = u$, $t \ge 0$; they satisfy the equation

$$Au + f(u) = 0. \tag{3.1.1}$$

We say that a solution u^* of Eq. (3.1.1), i.e., an equilibrium point u^*, is *hyperbolic* if the spectrum $\sigma(A + f'(u^*))$ is disjoint from the imaginary axis, i.e., $\sigma(A + f'(u^*)) \cap i\mathbb{R} = \varnothing$.

Proposition 3.1. *Assume that all equilibrium points of the problem (1.2.16) are hyperbolic. Then there is only a finite number of them and this number is odd.*

Proof. First, we note that, since $f(\cdot) : E^\theta \to E$ is bounded, all solutions of Eq. (3.1.1) satisfy the equation

$$Iu + A^{-1}f(u) = 0 \tag{3.1.2}$$

and hence $\|u\|_{E^1} \le \|f(u)\|_E \le C_1$. So, if we consider a ball of radius greater than $\|A^{-1}\|K$, then the operator $A^{-1}f(\cdot)$ maps the ball $\mathcal{U}_{E^\theta}(0, \|A^{-1}\|K) \subset E^\theta$ into itself. By the Schauder fixed-point theorem (see [168, Theorem 21.5]),

$$\gamma\Big(I + A^{-1}f(\cdot),\ \partial\mathcal{U}_{E^\theta}\big(0, \|A^{-1}\|K\big)\Big) = 1,$$

i.e., there exists at least one fixed point $Iu^* + A^{-1}f(u^*) = 0$ such that $u^* \in \mathcal{U}_{E^\theta}(0, \|A^{-1}\|K)$. Since the operator $A^{-1}f(\cdot) : E^\theta \to E^\theta$ is compact, we conclude that the set $\mathcal{E} = \{u : Au + f(u) = 0\}$ is compact in E^θ. Moreover, any fixed point u^* is isolated and $\big|\operatorname{ind}(u^*, I + A^{-1}f(\cdot))\big| = 1$ since

u^* is hyperbolic and $\mathcal{N}\big(I + A^{-1}f'(u^*)\big) = \{0\}$ (see [168, Theorem 21.6]). If the number of the fixed points is infinite, i.e., we have a sequence $\{u_i^*\}_{i=1}^{\infty}$, then the sequence $-A^{-1}f(u_i^*) = u_i^* \to u_\infty^*$ converges on some subsequence $i \in \mathbb{N}' \subseteq \mathbb{N}$, which contradicts the fact that the fixed point u_∞^* is isolated. So the number of the equilibrium points is finite. Now by [168, Theorem 20.6], we have

$$1 = \gamma\Big(I + A^{-1}f(\cdot),\ \partial \mathcal{U}_{E^\theta}\big(0, \|A^{-1}\|K\big)\Big) = \sum_{i=1}^{d} \operatorname{ind}\big(u_i^*, I + A^{-1}f(\cdot)\big)$$

and, therefore, $d = 2k + 1$ for some integer $k \geq 0$. $\qquad\square$

3.1.2 Unstable manifolds

Next we show that, in addition to stationary solutions, attractors contain many other solutions. It is easy to see that every mild solution of the problem (1.2.16) remains bounded on any $u^0 \in E^\theta$, i.e.,

$$\sup_{t \geq 0} \big\| u(t, u^0) \big\|_{E^\theta} < \infty.$$

If we show that the mild solution $u(\cdot, u^0)$ is defined on $\mathbb{R}$, takes value in E^θ, and is bounded, then it must belong to the attractor $\mathcal{A}$. Thus, to obtain a solution in the attractor, it suffices to show that a solution is defined and bounded in $\overline{\mathbb{R}}_- = (-\infty, 0]$. We introduce the unstable manifold

$$W^u(u^*) = \Big\{\eta \in E^\theta\ :\ u(t, \eta) \text{ is defined for all } t \leq 0$$

$$\text{and } u(t, \eta) \to u^* \text{ as } t \to -\infty\Big\}.$$

Before constructing $W^u(u^*)$ at least in a neighborhood of u^*, we briefly discuss linear problems. We naturally come to the linear case if we perform the change of variables $v(\cdot) = u(\cdot) - u^*$ in the problem (1.2.16). We obtain

$$\begin{aligned} v'(t) &= (A + f'(u^*))v(t) + f(v(t) + u^*) - f(u^*) - f'(u^*)v(t), \\ v(0) &= u^0 - u^* = v^0. \end{aligned} \tag{3.1.3}$$

In this equation, for very small v^0, the part $f(v(t)+u^*) - f(u^*) - f'(u^*)v(t)$ is also very small. Then it is natural to consider what happens when we neglect the nonlinearity; that is, what happens with the Cauchy problem

$$\begin{aligned} v'(t) &= (A + f'(u^*))v(t), \\ v(0) &= v^0. \end{aligned} \tag{3.1.4}$$

The operator $f'(u^*) \in B(E^\theta, E)$; therefore, the operator $A_{u^*} = A + f'(u^*)$ defined on $D(A)$ is a generator of an analytic C_0-semigroup with compact

resolvent (see [235]). It is well known that the spectrum of this generator consists just of isolated eigenvalues with finite-dimensional generalized
eigenspaces (the so-called discrete spectrum). The part σ^+ of the spectrum
of the operator $A + f'(u^*)$ lies to the right of the imaginary axis and consists
of a finite number of eigenvalues with finite multiplicities (see Fig. 3.1).

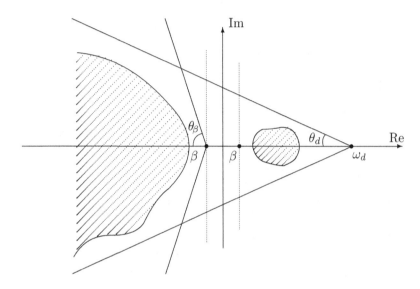

Fig. 3.1

Consider the decomposition $E^\theta = P(\sigma^+)E^\theta \oplus (I - P(\sigma^+))E^\theta$ using the
projector

$$P(\sigma^+) := P(\sigma^+, A_{u^*}) := \frac{1}{2\pi i} \int_{\partial U(\sigma^+)} (\zeta I - A_{u^*})^{-1} d\zeta \qquad (3.1.5)$$

defined by the set σ^+, where $U(\sigma^+)$ is a neighborhood of σ^+ such that
$U(\sigma^+) \cap i\mathbb{R} = \varnothing$. We see that there exists $\omega > 0$ such that

$$\begin{cases} \left\| e^{t(A+f'(u^*))} P(\sigma^+) \right\|_{B(E^\theta)} \leq M e^{\omega t}, & t \leq 0, \\ \left\| e^{t(A+f'(u^*))} (I - P(\sigma^+)) \right\|_{B(E^\theta, E^\phi)} \leq M t^{\theta-\phi} e^{-\omega t}, & \phi > \theta, \ t \geq 0. \end{cases}$$
$$(3.1.6)$$

If $P(\sigma^+)$ is the spectral projector defined by the part of the spectrum of
$(A+f'(u^*))$ lying to the right of the imaginary axis and $v_0 \in P(\sigma^+)E^\theta$, then
the solution $v(t, v_0)$ of (3.1.4) exists for all negative times t, and $v(t, v_0) \to 0$
as $t \to -\infty$; therefore, $v(t, v_0) + u^* \to u^*$ as $t \to -\infty$.

When we perturb (3.1.4) with a small nonlinearity, we observe solutions of (3.1.3) that exist for all negative time. Of course, the initial data for which such solutions exist, will no longer lie in $P(\sigma^+)E^\theta$, but in a nonlinear manifold near it (see Fig. 3.2).

Theorem 3.1 (see [69]). *Assume that u^* is a hyperbolic equilibrium for the problem (1.2.16), which is an isolated solution of Eq. (3.1.1). Then there exists a neighborhood U of u^* in E^θ such that the local unstable manifold $W^u(u^*)$ is given as follows:*

$$W^u_\delta(u^*) = \left\{ w(\cdot) \in W^u(u^*) : \|w(\cdot) - u^*\|_{E^\theta} < \delta \right\}$$
$$= \left\{ w(\cdot) : \left(v(\cdot), z(\cdot) \right) \in E^\theta : u \in U \right\}.$$

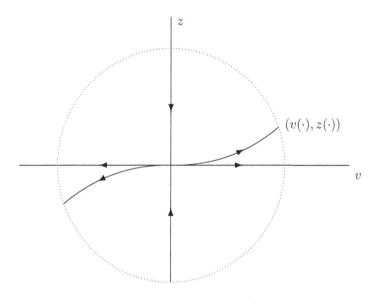

Fig. 3.2

Proof. Rewriting the problem (1.2.16) for $w(t) = u(t) - u^*$ to deal with the neighborhood of $u^* \in \mathcal{E}$, we arrive at the equation

$$w'(t) = (A + f'(u^*))w(t) + f(w(t) + u^*) - f(u^*) - f'(u^*)w(t). \quad (3.1.7)$$

If $w(\cdot)$ is a mild solution of (3.1.7), we write

$$v(t) = P(\sigma^+)w(t), \quad z(t) = (I - P(\sigma^+))w(t).$$

Hence, if B denotes the restriction of $(A + f'(u^*))$ to $P(\sigma^+)E^\theta$, then, denoting by $\tilde{A}$ the restriction of $(A + f'(u^*))$ to $(I - P(\sigma^+))E^\theta$, we obtain

$$
\begin{aligned}
v'(t) &= Bv(t) + P(\sigma^+)f\big(v(t) + z(t) + u^*\big) - P(\sigma^+)f(u^*) \\
&\quad - P(\sigma^+)f'(u^*)(v(t) + z(t)), \\
z'(t) &= \tilde{A}z(t) + \big(I - P(\sigma^+)\big)f\big(v(t) + z(t) + u^*\big) - \big(I - P(\sigma^+)\big)f(u^*) \\
&\quad - \big(I - P(\sigma^+)\big)f'(u^*)\big(v(t) + z(t)\big)
\end{aligned}
$$

(see Fig. 3.1 for the location of parts of the spectrum). We write

$$
\begin{aligned}
H(v, z) &= P(\sigma^+)f\big(v(t) + z(t) + u^*\big) - P(\sigma^+)f(u^*) \\
&\quad - P(\sigma^+)f'(u^*)\big(v(t) + z(t)\big), \\
G(v, z) &= \big(I - P(\sigma^+)\big)f\big(v(t) + z(t) + u^*\big) - \big(I - P(\sigma^+)\big)f(u^*) \\
&\quad - \big(I - P(\sigma^+)\big)f'(u^*)\big(v(t) + z(t)\big).
\end{aligned}
$$

Hence, we see that at the origin $(0,0)$, the functions H and G and their derivatives are equal to zero. Due to the continuous differentiability of H and G, we obtain that for given $\rho > 0$ and $\delta > 0$ such that

$$\|v\|_{E^\theta} + \|z\|_{E^\theta} < \delta,$$

we have

$$
\begin{cases}
\big\|H(v, z)\big\|_E \leq \rho, \\
\big\|G(v, z)\big\|_E \leq \rho, \\
\big\|H(v, z) - H(\tilde{v}, \tilde{z})\big\|_E \leq \rho\big(\|v - \tilde{v}\|_{E^\theta} + \|z - \tilde{z}\|_{E^\theta}\big), \\
\big\|G(v, z) - G(\tilde{v}, \tilde{z})\big\|_E \leq \rho\big(\|v - \tilde{v}\|_{E^\theta} + \|z - \tilde{z}\|_{E^\theta}\big).
\end{cases}
\tag{3.1.8}
$$

Using the above notation, we rewrite Eq. (3.1.7) in the following form:

$$
\begin{cases}
v'(t) = Bv(t) + H(v(t), z(t)), \\
z'(t) = \tilde{A}z(t) + G(v(t), z(t)),
\end{cases}
\tag{3.1.9}
$$

where H and G satisfy (3.1.8) for all

$$v(\cdot) \in P(\sigma^+)E^\theta, \quad z(\cdot) \in (I - P(\sigma^+))E^\theta.$$

Also, for some positive $M, \omega > 0$, due to the analyticity of the C_0-semigroup and splitting the space $E^\theta = P(\sigma^+)E^\theta \oplus (I - P(\sigma^+))E^\theta$, we have

$$
\begin{cases}
\big\|e^{\tilde{A}t}z\big\|_{E^\theta} \leq Me^{-\omega t}\|z\|_{E^\theta}, & t \geq 0, \\
\big\|e^{\tilde{A}t}z\big\|_{E^\theta} \leq Mt^{-\theta}e^{-\omega t}\|z\|_E, & t \geq 0, \\
\big\|e^{Bt}v\big\|_{E^\theta} \leq Me^{\omega t}\|v\|_E, & t \leq 0.
\end{cases}
\tag{3.1.10}
$$

Note that

$$\left\|e^{\tilde{A}t}z\right\|_{E^{\theta}} = \left\|(-A)^{\theta}e^{\tilde{A}t}z\right\| = \left\|(-A)^{\theta}\tilde{A}^{-\theta}(-\tilde{A})^{\theta}e^{\tilde{A}t}z\right\|;$$

since $(-A)^{\theta}\tilde{A}^{-\theta}$ is bounded, we obtain the inequalities (3.1.10). Here we also note that $P(\sigma^{+})E^{\theta}$ and $P(\sigma^{+})E$ are the same finite-dimensional subspaces of E and in this sense the E^{θ}-norm is equivalent to the E-norm in it, i.e.,

$$\|P(\sigma^{+})AP(\sigma^{+})\|_{B(E)} = \left\|\frac{1}{2\pi i}\int_{\Gamma}A(\lambda I - A)^{-1}d\lambda\right\| \leq \text{const.}$$

Now we show that for a suitably small $\rho > 0$, there exists a local unstable manifold for $u^{*} \in \mathcal{E}$

$$W^{u} = \left\{(v(\cdot), z(\cdot)) : z(\cdot) \in (I - P(\sigma^{+}))E^{\theta}, \ v(\cdot) \in P(\sigma^{+})E^{\theta}\right\}.$$

The solution $(v(t), z(t))$ must tend to zero as $t \to -\infty$ and, in particular, it must be bounded. Since

$$z(t) = e^{\tilde{A}(t-t_0)}z(t_0) + \int_{t_0}^{t}e^{\tilde{A}(t-s)}G(v(s), z(s))ds,$$

setting $t_0 \to -\infty$, we have

$$z(t) = \int_{-\infty}^{t}e^{\tilde{A}(t-s)}G(v(s), z(s))ds.$$

A mild solution of (3.1.9) is a vector-valued function $\zeta(t) = \begin{pmatrix} v(t) \\ z(t) \end{pmatrix}$ satisfying the equation

$$\zeta(t) = \Theta(\eta, \zeta(t)) \equiv \begin{pmatrix} e^{B(t-\tau)}\eta + \int_{\tau}^{t}e^{B(t-s)}H(v(s), z(s))ds \\ \int_{-\infty}^{t}e^{\tilde{A}(t-s)}G(v(s), z(s))ds \end{pmatrix}. \qquad (3.1.11)$$

Now we consider Eq. (3.1.11) in the space

$$\Upsilon = C_0\big((-\infty, \tau]; PE^{\theta}\big) \times C_0\big((-\infty, \tau]; (I - P)E^{\theta}\big)$$

equipped with the sup-norms under the condition $v(-\infty) = z(-\infty) = 0$. Let us show that $\Theta(\eta, \cdot) : \Upsilon \to \Upsilon$.

Take $\zeta(\cdot) \in \Upsilon$; then, according to (3.1.8), for any $\epsilon > 0$ there exists a number $-\infty < t_{\epsilon} < 0$ such that

$$\|G(\zeta(s))\|_{(I-P)E} \leq \epsilon, \quad \|H(\zeta(s))\|_{PE} \leq \epsilon, \quad s \leq t_{\epsilon}.$$

From (3.1.10) it follows that $e^{B(\cdot - \tau)}\eta \in C_0((-\infty, \tau]; PE^\theta)$ for any $\eta \in PE^\theta$ and

$$
\left\| (-A)^\theta \int_\tau^t e^{(t-s)B} H(\zeta(s)) ds \right\|
$$

$$
\leq C \left\| \int_\tau^t e^{(t-s)B} H(\zeta(s)) ds \right\|
$$

$$
\leq C \left\| e^{(t-t_\epsilon)A} \int_\tau^{t_\epsilon} e^{(t_\epsilon - s)B} H(\zeta(s)) ds + \int_{t_\epsilon}^t e^{(t-s)B} H(\zeta(s)) ds \right\|
$$

$$
\leq CM^2 e^{\omega(t-t_\epsilon)} \int_{t_\epsilon}^\tau e^{\omega(t_\epsilon - s)} K \, ds + M\epsilon \int_t^{t_\epsilon} e^{\omega(t-s)} ds
$$

$$
\leq CM^2 e^{\omega(t-t_\epsilon)} \frac{K}{\omega} + \epsilon \frac{M}{\omega}
$$

for $s \leq t_\epsilon$. Hence, for given $\epsilon > 0$,

$$
0 \leq \limsup_{t \to -\infty} \left\| (-A)^\theta \int_\tau^t e^{(t-s)B} H(\zeta(s)) ds \right\| \leq \epsilon \frac{M}{\omega}.
$$

This shows that the first coordinate of $\Theta(\eta, \zeta(t))$ lies in $C_0((-\infty, \tau]; PE^\theta)$. It is easy to see that the second coordinate of $\Theta(\eta, \zeta(t))$ lies in $C_0((-\infty, \tau]; (I - P)E^\theta)$.

If we set $\eta = \eta_0 = 0$ and $\zeta(t) = \zeta_0(t) = \begin{pmatrix} v_0(t) \\ z_0(t) \end{pmatrix} = 0$, $t \in (-\infty, \tau]$, then $\zeta_0(t) = \Theta(0, \zeta_0(t))$ and the operator $\Theta(\cdot, \cdot)$ is continuous in both arguments. Moreover, the Fréchet derivative $\Theta'_\zeta(\eta_0, \zeta_0) : \Upsilon \to \Upsilon$ is equal to

$$
\Theta'_\zeta(\eta_0, \zeta_0)h
$$

$$
= \begin{pmatrix} \int_\tau^t e^{B(t-s)} H'_v(\zeta_0(s))h^1(s) ds & \int_\tau^t e^{B(t-s)} H'_z(\zeta_0(s))h^2(s) ds \\ \int_{-\infty}^t e^{\tilde{A}(t-s)} G'_v(\zeta_0(s))h^1(s) ds & \int_{-\infty}^t e^{\tilde{A}(t-s)} G'_z(\zeta_0(s))h^2(s) ds \end{pmatrix} = 0
$$

for any $h = \begin{pmatrix} h^1(t) \\ h^2(t) \end{pmatrix} \in \Upsilon$ since the derivatives of the functions G and H vanish at the origin. Therefore, the operator $(I - \Theta'_\zeta(\eta_0, \zeta_0))^{-1}$ exists and is continuous. Due to [168, Theorem 54.2], we conclude that there are positive δ and ρ such that for $\|\eta - \eta_0\|_{QE^\theta} \leq \delta$, Eq. (3.1.11) has a unique solution $\zeta(\cdot)$ in the ball $\|\zeta - \zeta_0\|_\Upsilon \leq \rho$, which continuously depends on η.

This implies that the unstable manifold is the graph of the function

$$
\pounds : \{\eta \in PE^\theta : \|\eta\|_{E^\theta} \leq \delta\} \to (I - P)E^\theta
$$

defined by $\mathcal{L}(\eta) = z(\tau)$, where $z(\tau)$ is the second coordinate of the unique solution of Eq. (3.1.11). The above reasoning implies that $\mathcal{L}$ is a continuous function. □

Next, we assume that the problem (1.2.16) has gradient structure.

Definition 3.1. We say that $\{S(t) : t \geq 0\}$ is *gradient* if there is a continuous function $V : E^\theta \to \mathbb{R}$ such that the function $\overline{\mathbb{R}}^+ \ni t \mapsto V(S(t)u^0) \in \mathbb{R}$ is nonincreasing and

$$\mathcal{E} = \left\{ u \in E^\theta : Au + f(u) = 0 \right\} = \left\{ u^0 \in E^\theta : V(S(t)u^0) = V(u^0) \,\forall t \in \mathbb{R} \right\}.$$

Lemma 3.1. *Let B be a relatively compact invariant subset of E^θ under $\{S(t) : t \geq 0\}$. If $v \in B$, then there are $v^+, v^- \in \mathcal{E}$ such that $v \in W^u(v^+)$ and $v \in W^s(v^-)$.*

Proof. Let $v \in B$. Since B is invariant, there is a complete orbit $\mathbb{R} \ni t \mapsto S(t)v \in B$ passing through v. Since B is compact, this complete orbit is compact. The result now follows from the previous lemmas. □

This lemma implies the following result.

Theorem 3.2 (see [69]). *Let $\mathcal{A}$ be an attractor for a gradient semiflow $\{S(t) : t \geq 0\}$ with a compact resolvent $(\lambda I - A)^{-1}$ and let any point of $\mathcal{E}$ be hyperbolic. Then*

$$\mathcal{A} = \bigcup_{u^* \in \mathcal{E}} W^u(u^*).$$

3.2 Approximation of Operators on the General Discretization Scheme

For general examples of the $\mathcal{P}$-convergence, see [113, 271, 284, 289].

Theorem 3.3. *Let $B \in B(E)$ and $B_n \in B(E_n)$. The following conditions are equivalent:*

(i) $B_n \xrightarrow{\mathcal{PP}} B$ *as $n \to \infty$;*

(ii) $\|B_n\| \leq \text{const}, n \in \mathbb{N}$, *and* $\|B_n p_n x - p_n B x\| \to 0$ *as $n \to \infty$ for any $x \in E$;*

(iii) $\|B_n\| \leq \text{const}, n \in \mathbb{N}$, *and* $\|B_n x_n - p_n B x\| \to 0$ *as $n \to \infty$ for any $x \in E$ whenever $x_n \xrightarrow{\mathcal{P}} x \in E$.*

3.2.1   *Approximation of spectra of linear operators*

The most important role in approximations of the equation $Bx = y$ and approximations of the spectrum of an operator B is played by the notions of *stable* and *regular* convergence. These notions are used in various areas of numerical analysis (see [13, 112, 117, 118, 285, 288]).

Definition 3.2. We say that a sequence of operators $\{B_n\}$, $B_n \in B(E_n)$, $n \in \mathbb{N}$, *stably converges* to an operator $B \in B(E)$ if $B_n \xrightarrow{\mathcal{PP}} B$ and $\|B_n^{-1}\|_{B(E_n)} = O(1)$, $n \to \infty$. We denote this fact as follows: $B_n \xrightarrow{\mathcal{PP}} B$ stably.

Definition 3.3. We say that a sequence of operators $\{B_n\}$, $B_n \in B(E_n)$, *regularly converges* to an operator $B \in B(E)$ if $B_n \xrightarrow{\mathcal{PP}} B$ and the following implication holds:

$$\|x_n\|_{E_n} = O(1) \ \& \ \{B_n x_n\} \text{ is } \mathcal{P}\text{-compact} \implies \{x_n\} \text{ is } \mathcal{P}\text{-compact}.$$

We denote this fact as follows: $B_n \xrightarrow{\mathcal{PP}} B$ regularly.

Theorem 3.4 (see [288]). *For $B_n \in B(E_n)$ and $B \in B(E)$, the following conditions are equivalent*:

(i) $B_n \xrightarrow{\mathcal{PP}} B$ *regularly*, B_n *are Fredholm operators of index* 0, *and* $\mathcal{N}(B) = \{0\}$;

(ii) $B_n \xrightarrow{\mathcal{PP}} B$ *stably and* $\mathcal{R}(B) = E$;

(iii) $B_n \xrightarrow{\mathcal{PP}} B$ *stably and regularly*;

(iv) *if one of the conditions* (i)–(iii) *holds, then there exist* $B_n^{-1} \in B(E_n)$ *and* $B^{-1} \in B(E)$, *such that* $B_n^{-1} \xrightarrow{\mathcal{PP}} B^{-1}$ *regularly and stably*.

This theorem admits a generalization to the case of closed operators $B \in \mathcal{C}(E)$ and $B_n \in \mathcal{C}(E_n)$ (see [291]). Let $\Lambda \subseteq \mathbb{C}$ be an open connected set and let $B \in B(E)$. For an isolated point $\lambda \in \sigma(B)$, we denote by $\mathcal{W}(\lambda; B) = P(\lambda)E$ the corresponding maximal invariant space (or the generalized eigenspace), where

$$P(\lambda) = \frac{1}{2\pi i} \int_{|\zeta - \lambda| = \delta} (\zeta I - B)^{-1} d\zeta$$

is the Riesz projector and δ is sufficiently small so that there are no points of $\sigma(B)$ different from λ in the disc $\{\zeta : |\zeta - \lambda| \leq \delta\}$. An isolated point

$\lambda \in \sigma(B)$ is called a *Riesz point* of B if $\lambda I - B$ is a Fredholm operator of zero index and $P(\lambda)$ has finite rank. Let

$$W(\lambda, \delta; B_n) = \bigcup_{\substack{|\lambda_n - \lambda| < \delta, \\ \lambda_n \in \sigma(B_n)}} W(\lambda_n; B_n),$$

where $\lambda_n \in \sigma(B_n)$ are taken from a δ-neighborhood of λ. It is clear that $W(\lambda, \delta; B_n) = P_n(\lambda) E_n$, where

$$P_n(\lambda) = \frac{1}{2\pi i} \int_{|\zeta - \lambda| = \delta} (\zeta I - B_n)^{-1} d\zeta.$$

The following theorems state a complete picture of the approximation of the spectrum.

Theorem 3.5 (see [113, 286, 287]). *Assume that $L_n(\lambda) = \lambda I - B_n$ and $L(\lambda) = \lambda I - B$ are Fredholm operators of zero index for any $\lambda \in \Lambda$ and $L_n(\lambda) \to L(\lambda)$ stably for any $\lambda \in \rho(B) \cap \Lambda \neq \varnothing$. Then the following assertions hold:*

(i) *for any $\lambda_0 \in \sigma(B) \cap \Lambda$, there exists a sequence $\{\lambda_n\} \subset \sigma(B_n)$, $n \in \mathbb{N}$, such that $\lambda_n \to \lambda_0$ as $n \to \infty$;*
(ii) *if for some sequence $\{\lambda_n\} \subset \sigma(B_n)$, $n \in \mathbb{N}$, we have $\lambda_n \to \lambda_0 \in \Lambda$ as $n \to \infty$, then $\lambda_0 \in \sigma(B)$;*
(iii) *for any $x \in W(\lambda_0; B)$, there exists a sequence $\{x_n\} \subset W(\lambda_0, \delta; B_n)$, $n \in \mathbb{N}$, such that $x_n \to x$ as $n \to \infty$;*
(iv) *there exists $n_0 \in \mathbb{N}$ such that $\dim W(\lambda_0, \delta; B_n) \geq \dim W(\lambda_0; B)$ for any $n \geq n_0$.*

Remark 3.1. It was shown in [285] that the inequality in (iv) can be strict for all $n \in \mathbb{N}$.

Theorem 3.6 (see [285]). *Assume that $L_n(\lambda)$ and $L(\lambda)$ are Fredholm operators of zero index for all $\lambda \in \Lambda$ and $L_n(\lambda) \xrightarrow{\mathcal{PP}} L(\lambda)$ regularly for any $\lambda \in \Lambda$ and $\rho(B) \cap \Lambda \neq \varnothing$. Then the assertions (i)–(iii) of Theorem 3.5 hold and also*

(iv') *there exists $n_0 \in \mathbb{N}$ such that $\dim W(\lambda_0, \delta; B_n) = \dim W(\lambda_0; B)$ for all $n \geq n_0$;*
(v) *any sequence $\{x_n\}$, $x_n \in W(\lambda_0, \delta; B_n)$, $n \in \mathbb{N}$, such that $\|x_n\|_{E_n} = 1$ is $\mathcal{P}$-compact and any limit point of this sequence belongs to $W(\lambda_0; B)$.*

3.2.2   *Regions of convergence*

Theorems 3.5 and 3.6 were generalized to the case of closed operators in [291]. Now we use the following notions introduced by T. Kato (see [146]).

Definition 3.4. The *region of stability* $\Delta_s = \Delta_s(\{A_n\})$, $A_n \in \mathcal{C}(E_n)$, is defined as the set of all $\lambda \in \mathbb{C}$ such that $\lambda \in \rho(A_n)$ for almost all n and such that the sequence $\{\|(\lambda I - A_n)^{-1}\|\}_{n \in \mathbb{N}}$ is bounded. The *region of convergence* $\Delta_c = \Delta_c(\{A_n\})$, $A_n \in \mathcal{C}(E_n)$, is defined as the set of all $\lambda \in \mathbb{C}$ such that $\lambda \in \Delta_s(\{A_n\})$ and the sequence of operators $\{(\lambda I - A_n)^{-1}\}_{n \in \mathbb{N}}$ is $\mathcal{PP}$-convergent to some operator $S(\lambda) \in B(E)$.

Definition 3.5. A sequence of operators $\{K_n\}$, $K_n \in \mathcal{C}(E_n)$, is said to be *regularly compatible* with an operator $K \in \mathcal{C}(E)$ if (K_n, K) are compatible and for any bounded sequence $\|x_n\|_{E_n} = O(1)$ such that $x_n \in D(K_n)$ and $\{K_n x_n\}$ is $\mathcal{P}$-compact, we have that $\{x_n\}$ is $\mathcal{P}$-compact and the $\mathcal{P}$-convergence of $\{x_n\}$ to some x and the $\mathcal{P}$-convergence of $\{K_n x_n\}$ to some y as $n \to \infty$ in $\mathbb{N}' \subseteq \mathbb{N}$ imply that $x \in D(K)$ and $Kx = y$.

Definition 3.6. The *region of regularity* $\Delta_r = \Delta_r(\{A_n\}, A)$ is defined as the set of all $\lambda \in \mathbb{C}$ such that (K_n, K), where $K_n = \lambda I - A_n$ and $K = \lambda I - A$ are regularly compatible.

The relationships between these notions are given by the following assertion.

Proposition 3.2 (see [291]). *Assume that $\Delta_c \neq \varnothing$ and $\mathcal{N}(S(\lambda)) = \{0\}$ at least for one point $\lambda \in \Delta_c$, so that $S(\lambda) = (\lambda I - A)^{-1}$. Then (A_n, A) are compatible and*

$$\Delta_c = \Delta_s \cap \rho(A) = \Delta_s \cap \Delta_r = \Delta_r \cap \rho(A).$$

It was shown in [291] that the conditions "A_n and A are compatible," "$\lambda I_n - A_n$ and $\lambda I - A$ are Fredholm operators with zero index for any $\lambda \in \Lambda$," and "$\rho(A) \cap \Lambda \neq \varnothing$" imply Conditions (i)–(iv) of Theorem 3.5 if $\rho(A) \cap \Lambda \subseteq \Delta_s$ and imply Conditions (i)–(iii) of Theorem 3.5 and Conditions (iv)–(v) of Theorem 3.6 when $\Lambda \subseteq \Delta_r$.

Definition 3.7. A Riesz point $\lambda_0 \in \sigma(A)$ is said to be *strongly stable in the Kato sense* if $\dim \mathcal{W}(\lambda_0, \delta; A_n) \leq \dim \mathcal{W}(\lambda_0; A)$ for all $n \geq n_0$.

Theorem 3.7 (see [291]). *A Riesz point $\lambda_0 \in \sigma(A)$ is strongly stable in the Kato sense if and only if $\lambda_0 \in \Lambda \cap \Delta_r \cap \sigma(A)$.*

Definition 3.8. The region of compact convergence of the resolvents Δ_{cc}, where $A \in \mathcal{C}(E_n)$, $A \in \mathcal{C}(E)$, is defined as the set of all $\lambda \in \Delta_c \cap \rho(A)$ such that $(\lambda I_n - A_n)^{-1} \xrightarrow{PP} (\lambda I - A)^{-1}$ compactly.

Corollary 3.1. *Assume that* $\Delta_{cc} \neq \varnothing$. *Then* $\Delta_{cc} = \Delta_c \cap \rho(A)$.

Proof. It is clear that $\Delta_{cc} \subseteq \Delta_c \cap \rho(A)$. To prove that $\Delta_{cc} \supseteq \Delta_c \cap \rho(A)$, we consider the Hilbert identity (2.3.4). Now let $\mu \in \Delta_{cc}$. Then $\mu \in \Delta_{cc} \cap \Delta_c \cap \rho(A)$. Hence, for every $\zeta \in \Delta_c \cap \rho(A)$ and any bounded sequence $\{x_n\}$, $n \in \mathbb{N}$, the sequence $\{(\zeta I_n - A_n)^{-1} x_n\}$ is $\mathcal{P}$-compact. $\square$

Theorem 3.8. *Assume that* $\Delta_{cc} \neq \varnothing$. *Then* $\Delta_r = \mathbb{C}$.

Proof. Take an arbitrary point $\lambda_1 \in \mathbb{C}$. We must prove that $\lambda_1 I - A_n$ and $\lambda_1 I - A$ are regularly compatible. Assume that $\|x_n\|_{E_n} = O(1)$ and $\{(\lambda_1 I_n - A_n) x_n\}$ is $\mathcal{P}$-compact. To show that $\{x_n\}$ is $\mathcal{P}$-compact, we take $\mu \in \Delta_{cc}$. Using (2.3.4) with $\zeta = \lambda_1$, we obtain

$$x_n = (\mu I_n - A_n)^{-1}(\lambda_1 I_n - A_n) x_n + (\lambda_1 - \mu)(\mu I_n - A_n)^{-1} x_n$$

and, therefore, $\{x_n\}$ is $\mathcal{P}$-compact. Now assume that

$$x_n \xrightarrow{PP} x \quad \text{and} \quad (\lambda_1 I_n - A_n)^{-1} x_n \xrightarrow{P} y \quad \text{as } n \to \infty \text{ in } \mathbb{N}' \subseteq \mathbb{N}.$$

Then

$$x = (\mu I - A)^{-1} y - (\lambda_1 - \mu)(\mu I - A)^{-1} x,$$

which implies that $x \in D(A)$ and $(\lambda_1 I - A)x = y$. $\square$

Theorem 3.8 implies that in the case $\Delta_{cc} \neq \varnothing$, all assumptions of Theorem 3.6 are fulfilled.

Similarly to the operator $\Theta(\eta, u)$ (see (3.1.11)), we introduce the operator $\Theta_n(\eta_n, u_n)$. Normally we assume that $\eta_n \xrightarrow{\mathcal{P}^\theta} \eta$ or even $\eta_n = p_n^\theta \eta$. Rewriting (1.4.5) for $w_n(t) = u_n(t) - u_n^*$ to deal with a neighborhood of u_n^*, we arrive at the equation

$$\begin{aligned} w_n'(t) = A_n w_n(t) &+ f_n'(u_n^*) w_n(t) + f_n\big(w_n(t) + u_n^*\big) \\ &- f_n\big(u_n^*(t)\big) - f_n'(u_n^*) w_n(t). \end{aligned} \tag{3.2.1}$$

If P_n is the projector determined by the spectral set

$$\sigma_n^+ = \big\{\lambda \in \sigma(A_n + f_n'(u_n^*)) : \operatorname{Re}\lambda > 0\big\}$$

and $W_n = P_n E_n^\theta$, then W_n is isomorphic to $W = PE^\theta$ through the isomorphism $\Xi_n = P_n p_n^\theta$, which is bounded with bounded inverse Ξ_n^{-1}, and the norms of Ξ_n and Ξ_n^{-1} are uniformly bounded $n \geq n_0$. Recall that P_n are projectors such that $P_n \xrightarrow{\;\mathcal{P}^\theta\;} P(\sigma^+)$ compactly (for the definition of $P(\sigma^+)$, see in Sec. 3.1.2). We identify all spaces W_n and $P_n E_n$ through these isomorphisms to the fixed space W.

Now we decompose Eq. (3.2.1) as follows. If w_n is a solution of (3.2.1), we write

$$w_n(t) = v_n(t) + z_n(t),$$

where

$$v_n(t) = P_n w_n(t), \quad z_n(t) = (I_n - P_n)w_n(t).$$

Therefore, if we denote by B_n the bounded linear operator given as the restriction of A_n to $P_n E_n$, we have

$$v_n' = B_n v_n(t) + P_n\Big(f_n(v_n + z_n + u_n^*) - f_n(u_n^*) - f_n'(u_n^*)(v_n + z_n) \Big),$$

$$z_n' = A_n z_n + f_n'(u_n^*)z$$
$$+ (I_n - P_n)\Big(f_n(v_n + z_n + u_n^*) - f_n(u_n^*) - f_n'(u_n^*)(v_n + z_n) \Big).$$

We write

$$H_n(v_n, z_n) = P_n\Big(f_n(v_n + z_n + u_n^*) - f_n(u_n^*) - f_n'(u_n^*)(v_n + z_n) \Big),$$

$$G_n(v_n, z_n) = (I_n - P_n)\Big(f_n(v_n + z_n + u_n^*) - f_n(u_n^*) - f_n'(u_n^*)(v_n + z_n) \Big).$$

Hence, we have $H_n(0,0) = 0$ and $G_n(0,0) = 0$. The continuous differentiability of H_n and G_n and the estimates from Sec. 3.5, which hold uniformly in n, imply that for given $\rho > 0$, there exist $n_0 > 0$ and $\delta > 0$ such that if $\|v_n\|_{W_n} + \|z_n\|_{E_n^\theta} < \delta$ and $n \geq n_0$, then

$$\begin{cases} \big\|H_n(v_n, z_n)\big\|_{W_n} \leq \rho, \\[4pt] \big\|G_n(v_n, z_n)\big\|_{E_n} \leq \rho, \\[4pt] \big\|H_n(v_n, z_n) - H_n(\tilde{v}_n, \tilde{z}_n)\big\|_{W_n} \leq \rho\big(\|v_n - \tilde{v}_n\|_{W_n} + \|z_n - \tilde{z}_n\|_{E_n^\theta}\big), \\[4pt] \big\|G_n(v_n, z_n) - G_n(\tilde{v}_n, \tilde{z}_n)\big\|_{E_n} \leq \rho\big(\|v_n - \tilde{v}_n\|_{W_n} + \|z_n - \tilde{z}_n\|_{E_n^\theta}\big). \end{cases}$$
$$(3.2.2)$$

The fact that we can choose ρ and δ, which satisfy the inequalities above uniformly for $n \geq n_0$, is the key point to obtain that local unstable manifolds are defined in a small "neighborhood" of the equilibrium point u_n uniformly for $n \geq n_0$.

Let $\tilde{A}_n = (A_n + f'_n(u^*_n))\big|_{(I_n - P_n)E^\theta_n}$. Then Eq. (3.2.1) can be rewritten as follows:

$$\begin{cases} v'_n(t) = \big(A_n + f'_n(u^*_n)\big)P_n v_n(t) + H_n\big(v_n(t), z_n(t)\big), \\ z'_n(t) = \big(A_n + f'_n(u^*_n)\big)(I_n - P_n)z_n(t) + G_n\big(v_n(t), z_n(t)\big), \end{cases} \quad (3.2.3)$$

where H_n and G_n satisfy (3.2.2) for all $v_n(\cdot) \in W_n$ and $z_n(\cdot) \in (I_n - P_n)E^\theta_n$.

Let $M, \omega > 0$ and

$$\Omega^{M,\omega} = \Big\{ (v(\cdot), z(\cdot)) \in C_0\big((-\infty, 0]; PE^\theta\big) \times C_0\big((-\infty, 0]; (I - P)E^\theta\big) :$$

$$\|v(t)\|_{E^\theta}, \; \|z(t)\|_{E^\theta} \le Me^{\omega t}, \; t \le 0 \Big\}.$$

Theorem 3.9. *The operator* $\Theta(\eta, \cdot) : \Omega^{M,\omega} \subset \Upsilon \to \Upsilon$ *is compact.*

Proof. Consider sequences of bounded functions $\{v^k(\cdot)\}$ and $\{z^k(\cdot)\}$, $v^k, z^k \in \Omega^{M,\omega}$, $k \in \mathbb{N}$. Then the operators

$$(-A)^\theta \int_0^t e^{(t-s)(A+f'(u^*))} QH(v^k(s), z^k(s))ds,$$

$$(-A)^\theta \int_{-\infty}^t e^{(t-s)(A+f'(u^*))}(I - Q)G(v^k(s), z^k(s))ds$$

map these sequences into compact sequences. Indeed, one can easily verify the uniform boundedness and the compactness for any fixed $t < 0$. To verify the equicontinuity, note that

$$\left\| (-A)^\theta \int_0^t e^{(t-s)(A+f'(u^*))} QH(v^k(s), z^k(s))ds \right\| \le \epsilon$$

whenever $\|v^k(s)\|, \|z^k(s)\| \le \delta$ for $t \le t_\epsilon$. Thus, applying the Arzelà–Ascoli theorem on the interval $[t_\epsilon, 0]$, we obtain the equicontinuity (cf. Theorem 1.6). $\square$

Theorem 3.10. *The operators*

$$\Theta_n(\eta_n, \cdot) : \Omega^{M,\omega}_n \subset \Upsilon_n \to \Upsilon_n, \quad \Theta(\eta, \cdot) : \Omega^{M,\omega} \subset \Upsilon \to \Upsilon$$

converge compactly: $\Theta_n(\eta_n, \cdot) \to \Theta(\eta, \cdot)$.

Proof. Recall that

$$\|u_n(\cdot)\|_{\Upsilon_n} = \big\|(u^1_n(\cdot), u^2_n(\cdot))\big\|_{\Upsilon_n}$$

$$= \sup_{-\infty < t \le 0} \big\|u^1_n(t)\big\|_{P_n E^\theta_n} + \sup_{-\infty < t \le 0} \big\|u^2_n(t)\big\|_{(I_n - P_n)E^\theta_n}$$

and $u_n^1(-\infty) = u_n^2(-\infty) = 0$. Thus, the discrete convergence

$$\Upsilon_n \ni u_n(\cdot) \to u(\cdot) \in \Upsilon$$

means that

$$\sup_{-\infty < t \leq 0} \left\| u_n^1(t) - p_n^\theta u^1(t) \right\|_{P_n E_n^\theta} + \sup_{-\infty < t \leq 0} \left\| u_n^2(t) - p_n^\theta u^2(t) \right\|_{(I_n - P_n) E_n^\theta} \to 0$$

as $n \to \infty$. Verify that the condition $\|u_n(\cdot)\|_{\Upsilon_n} = O(1)$ implies that the sequence $\{\Theta_n(\eta_n, u_n(\cdot))\}$ is compact. We apply the Arzelà–Ascoli theorem to the sequences

$$\left\{ (-A_n)^\theta \int_0^t e^{(t-s)\left(A_n + f_n'(u_n^*)\right)} P_n H_n\big(v_n^k(s), z_n^k(s)\big) ds \right\},$$

$$\left\{ (-A_n)^\theta \int_{-\infty}^t e^{(t-s)\left(A_n + f_n'(u_n^*)\right)} (I_n - P_n) G_n\big(v_n^k(s), z_n^k(s)\big) ds \right\}.$$

Due to (3.2.2), we have

$$\left\| (-A_n)^\theta \int_0^t e^{(t-s)A_n} H_n\big(v_n^k(s), z_n^k(s)\big) ds \right\| \leq \epsilon$$

for $t \leq t_\epsilon$. Now we apply the Arzelà–Ascoli theorem to the sequences of functions

$$\left\{ (-A_n)^\theta \int_0^t e^{(t-s)(A_n + f_n'(u_n^*))} P_n H_n\big(v_n^k(s), z_n^k(s)\big) ds \right\},$$

$$\left\{ (-A_n)^\theta \int_{-\infty}^t e^{(t-s)(A_n + f_n'(u_n^*))} (I_n - P_n) G_n\big(v_n^k(s), z_n^k(s)\big) ds \right\}$$

just on $[t_\epsilon, 0]$. This can be done as in Theorem 1.6.   □

3.3   Approximation of Attractors

Next we intend to obtain the convergence of attractors. We use Definition 2.1 to formulate the statements of Lemma 2.1. These conclusions will be important to operate with the upper and lower semicontinuity.

3.3.1   *Upper semicontinuity of attractors*

In this subsection, we obtain that the family of attractors $\{\mathcal{A}_n\}$ behaves upper-semicontinuously. In order to prove this, we use the results on the continuity of nonlinear semiflows and the following lemma.

As before, we assume that the operators A_n generate analytic C_0-semigroups satisfying the estimate

$$\left\| e^{tA_n} \right\| \leq M e^{-t\tilde{\omega}}, \quad t \geq 0,$$

where $\tilde{\omega} > 0$ and the resolvents $(\lambda I_n - A_n)^{-1}$ are compact for some $\lambda \in \rho(A_n)$ and converge compactly, $(\lambda I_n - A_n)^{-1} \to (\lambda I - A)^{-1}$, for some $\lambda \in \bigcap_{n=1}^{\infty} \rho(A_n) \cap \rho(A)$. The functions $f_n(\cdot) : E_n^{\theta} \to E_n$ are globally Lipschitzian, bounded (i.e., there is a constant $\tilde{K} > 0$ such that $\|f_n(w)\|_E \leq \tilde{K}$ for all $w_n \in E_n^{\theta}$), and are continuously Fréchet differentiable functions.

Lemma 3.2. *For given $\theta \leq \gamma < 1$, there exists a constant N_{γ} independent of n such that*

$$\sup_{n \in \mathbb{N}} \sup_{u_n \in \mathcal{A}_n} \|u_n\|_{E_n^{\gamma}} \leq N_{\gamma}.$$

Moreover, any sequence $\{u_n\}$ with $u_n \in \mathcal{A}_n$ has a $\mathcal{P}$-convergent subsequence.

Proof. The first assertion of the lemma follows immediately from the invariance of the attractor and the formula of variation of constants. For the second assertion, we only need to apply Lemma 3.7. $\square$

Theorem 3.11. *Let $\mathcal{A}_n$ and $\mathcal{A}$ be the attractors for the problems (1.2.16) and (1.4.5), respectively. If the sequence $\{x_n^*\}$, $x_n^* \in \mathcal{A}_n$, $\mathcal{P}^{\theta}$-converges to some point x^*, then $x^* \in \mathcal{A}$ and $\{\mathcal{A}_n\}$ is $\mathcal{P}^{\theta}$-upper semicontinuous at infinity.*

Proof. Assume that $x_n^* \in \mathcal{E}_n$. In this case, by Proposition 3.3, we have $x^* \in \mathcal{A}$. Now assume that x_n^* does not belong to $\mathcal{E}_n$. In this case, x_n^* can converge to some point in $\mathcal{E}$ and there is nothing to prove. Let us consider the case where the limit point $x_n^* \xrightarrow{\mathcal{P}} x$ does not lie in $\mathcal{E}$. In this situation, consider back-time trajectories starting from these points as $u_n(0) = x_n^*$ and $u(0) = x$. On the interval $(-\infty, 0]$, the complete trajectories $u_n(t)$, $t \in (-\infty, 0]$, must be uniformly bounded in n and t since the functions $f_n(\cdot)$ are uniformly bounded in n and all attractors $\mathcal{A}_n$ are uniformly bounded in n. By Lemma 3.2, any sequence $\{y_n\}$ with $y_n \in \mathcal{A}_n$, $n \in \mathbb{N}$, is discretely compact. Now taking bounded trajectories $u_n(\cdot)$ passing through y_n and considering the sequence $\{u_n(-1)\}$, we obtain that $\{u_n(-1)\}$ is also discretely compact. Taking a subsequence $n \in \mathbb{N}' \subset \mathbb{N}$, we see that

$$u_n(-1) \xrightarrow{\mathcal{P}^{\theta}} \tilde{u}(-1) \quad \text{as } n \to \infty, \, n \in \mathbb{N}'.$$

It follows that

$$S_n(t)u_n(-1) \to S(t)\tilde{u}(-1) \quad \text{for any } t \in [0, T],$$

and, in particular,

$$S_n(1)u_n(-1) = x_n^* \xrightarrow{\mathcal{P}} S(1)\tilde{u}(-1) = x.$$

Now, taking the sequence $u_n(-2)$, we conclude that the trajectory $S(t)\tilde{u}(-2)$ is bounded for $t \geq 0$ and coincide with the trajectory $S(t)\tilde{u}(-1)$ for $t \geq 0$ for common t. In this way, we can construct a trajectory $u(t)$, $t \in \mathbb{R}$, which is bounded and contains the point x. Thus, $x \in \mathcal{A}$. $\qquad\square$

3.3.2 *Approximation of the set of equilibria*

In Banach spaces E_n, consider the following family of nonlinear problems:

$$A_n u_n + f_n(u_n) = 0. \tag{3.3.1}$$

Proposition 3.3. *Assume that $\Delta_{cc} \neq \varnothing$ and the problems* (3.3.1) *have solutions* $\{u_n^*\}$, $n \in \mathbb{N}$. *Then, passing to subsequences if necessary, we see that there is a solution u^* of Eq.* (3.1.1) *such that*

$$\left\| u_n^* - p_n^\theta u^* \right\|_{E_n^\theta} \to 0 \quad \text{as } n \to \infty.$$

Proof. Clearly,

$$u_n^* + A_n^{-1} f_n(u_n^*) = 0.$$

Since $A_n^{-1} \xrightarrow{\mathcal{PP}} A^{-1}$ compactly by Lemma 3.7 and $f_n(\cdot)$ are bounded uniformly in $n \in \mathbb{N}$, we conclude that there is a subsequence (we denote it again by u_n^*) and a point u^* such that $u_n^* \xrightarrow{\mathcal{P}^\theta} u^*$. The continuity of $f_n(\cdot)$ and the convergence of the resolvents $A_n^{-1} \xrightarrow{\mathcal{PP}} A^{-1}$ imply that $u^* + A^{-1} f^*(u^*) = 0$, which is equivalent to the fact that u^* is a solution of Eq. (3.1.1). $\qquad\square$

Proposition 3.4. *Let $A_n^{-1} \xrightarrow{\mathcal{PP}} A^{-1}$ compactly and $f_n(x_n) \xrightarrow{\mathcal{P}} f(x)$ as $x_n \xrightarrow{\mathcal{P}^\theta} x$. Assume also that u^* is a hyperbolic solution of Eq.* (3.1.1). *Then there exists n_0 such that for $n \geq n_0$, the problems* (3.3.1) *have at least one sequence of solutions $\{u_n^*\}$. Any sequence $\{u_n^*\}$, $n \in \mathbb{N}$, is $\mathcal{P}^\theta$-compact and the $\mathcal{P}^\theta$-limit point of $\{u_n^*\}$, $n \in \mathbb{N}$, is a solution u^* of Eq.* (3.1.1), *i.e.,*

$$u_n^* \xrightarrow{\mathcal{P}^\theta} u^* \text{ as } n \in \mathbb{N}' \subset \mathbb{N}.$$

Proof. As in Theorem 3.1, there exists a ball $\mathcal{U}(u^*, \delta)$ such that there are no other fixed points in it except for u^*, and we have

$$\left| \mathrm{ind}(u^*, I + A^{-1} f(\cdot)) \right| = 1.$$

The compact convergence

$$A_n^{-1} f_n(\cdot) \xrightarrow{\mathcal{PP}} A^{-1} f(\cdot)$$

follows from the boundedness and the continuity of $f_n(\cdot)$ and the compact convergence $A_n^{-1} \xrightarrow{\mathcal{PP}} A^{-1}$. Now the compact convergence

$$A_n^{-1} f_n(\cdot) \to A^{-1} f(\cdot)$$

implies (see [288, Theorem 3]) that

$$\text{ind}\left(u^*, I + A^{-1}f(\cdot)\right) = \gamma\left(I_n + A_n^{-1}f_n(\cdot),\ \partial \mathcal{U}(p_n u^*, \delta)\right), \quad n \geq n_0,$$

which means that in any ball $\mathcal{U}(p_n u^*, \delta)$, $n \geq n_0$, there exists at least one fixed point u_n^*. This sequence $\{u_n^*\}$ is $\mathcal{P}^\theta$-compact and converges on subsequences to u^*. $\qquad\square$

In the case where we have information on the differentiability of the corresponding functions, we have a more accurate result.

Proposition 3.5. *Assume that Eq. (3.1.1) has a solution u^*, which is a hyperbolic point. Assume also that in balls around the points $p_n^\theta u^*$, there exist the Fréchet derivatives of the functions $f_n(\cdot)$ and $f_n'(u_n)$ are uniformly continuous with respect to n in the balls $\mathcal{U}(p_n^\theta u^*, \delta)$ for some $\delta > 0$. Then there exists n_0 and $\delta > 0$ such that the problems (3.3.1) have exactly one solution u_n^* in $\{w_n : \|w_n - p_n^\theta u^*\|_{E_n^\theta} \leq \delta\}$ for $n \geq n_0$. Moreover,*

$$\left\|u_n^* - p_n^\theta u^*\right\|_{E_n^\theta} \to 0 \quad \text{as } n \to \infty.$$

Proof. We prove that in a neighborhood of a solution u^* of Eq. (3.1.1), there is a unique solution u_n^* of (3.3.1). Moreover, since $f_n'(\cdot)$ is uniformly continuous with respect to n, there is $\delta > 0$ such that the inequality

$$\left\|u_n^* - p_n^\theta u^*\right\|_{E_n^\theta} < \delta$$

implies

$$M\left\|f_n'(u_n^*) - f_n'(p_n^\theta u^*)\right\|_{B(E_n^\theta, E_n)} < \frac{1}{2C}.$$

By Lemma 3.8, $\sigma\left(A_n + f_n'(p_n^\theta u^*)\right)$ does not intersect the imaginary axis and there is a constant $M > 0$ such that

$$\left\|\left(A_n + f_n'(p_n^\theta u^*)\right)^{-1}\right\|_{B(E_n, E_n^\theta)} \leq M.$$

If u_n is a solution of (3.3.1), then

$$0 = A_n u_n + f_n'(p_n^\theta u^*)u_n + f_n(u_n) - f_n'(p_n^\theta u^*)u_n$$

$$= \left(A_n + f_n'(p_n^\theta u^*)\right)\left(u_n + \left(A_n + f_n'(p_n^\theta u^*)\right)^{-1}\left(f_n(u_n) - f_n'(p_n^\theta u^*)u_n\right)\right).$$

Since $\left(A_n + f_n'(p_n^\theta u^*)\right)$ is invertible, the proof of the fact that u_n is a solution of (3.3.1) is equivalent to the proof of the fact that it is a fixed point of the mapping

$$\Phi_n(u_n) = -\left(A_n + f_n'(p_n^\theta u^*)\right)^{-1}\left(f_n(u_n) - f_n'(p_n^\theta u^*)u_n\right).$$

It is clear from the convergence of the resolvent and the continuity of $f_n(\cdot)$ and $f_n'(p_n^\theta u^*)$ that

$$\Phi_n(p_n^\theta u^*) \to -\left(A + f'(u^*)\right)^{-1}\left(f(u^*) - f'(u^*)u^*\right) = u^*. \qquad (3.3.2)$$

Next we show that for sufficiently small $\delta > 0$ and $n \geq n_0$, the mapping Φ_n is a contraction from the closed ball $\bar{\mathcal{U}}(p_n u^*, \delta)$ into itself. First, we verify that Φ_n is a contraction (uniformly in $n \in \mathbb{N}$), i.e.,

$$\left\|\Phi_n(u_n) - \Phi_n(v_n)\right\|_{E_n^\theta}$$

$$= \left\|\left(A_n + f_n'(p_n^\theta u^*)\right)^{-1}\left(f_n(u_n) - f_n(v_n) - f_n'(p_n^\theta u^*)(u_n - v_n)\right)\right\|_{E_n^\theta}$$

$$\leq \left\|\left(A_n + f_n'(p_n^\theta u^*)\right)^{-1}\right\|_{L(E_n, E_n^\theta)}\left\|f_n(u_n) - f(v_n) - f_n'(p_n^\theta u^*)(u_n - v_n)\right\|_{E_n}$$

$$\leq M\left\|\left(f_n'(su_n + (1-s)v_n) - f_n'(p_n u^*)\right)(u_n - v_n)\right\|_{E_n}$$

$$\leq M\left\|f_n'(su_n + (1-s)v_n) - f_n'(p_n^\theta u^*)\right\|_{L(E_n^\theta, E_n)}\|u_n - v_n\|_{E_n^\theta}$$

$$\leq \frac{1}{2C}\|u_n - v_n\|_{E_n^\theta}$$

for some $0 \leq s \leq 1$ and $n \geq n_0$. To show that Φ_n maps $\mathcal{U}(p_n u^*, \delta)$ into itself, we note that if $u_n \in \mathcal{U}(p_n u^*, \delta)$, then

$$\left\|\Phi_n(u_n) - p_n u^*\right\|_{E_n^\theta} \leq \left\|\Phi_n(u_n) - \Phi_n(p_n u^*)\right\|_{E_n^\theta} + \left\|\Phi_n(p_n u^*) - p_n u^*\right\|_{E_n^\theta}$$

$$\leq \frac{\delta}{2} + \left\|\Phi_n(p_n u^*) - p_n u^*\right\|_{E_n^\theta}.$$

It follows from (3.3.2) that there exists n_0 such that

$$\left\|\Phi_n(u_n) - p_n u^*\right\|_{E_n^\theta} \leq \frac{3\delta}{4} \quad \text{for } n \geq n_0.$$

This implies that if $u_n \in \bar{\mathcal{U}}(p_n^\theta u^*, \delta)$, then

$$\left\|\Phi_n(u_n) - p_n^\theta u^*\right\|_{E_n^\theta} \leq \delta$$

and, therefore, $\Phi_n : \mathcal{U}(p_n u^*, \delta) \to \mathcal{U}(p_n u^*, \delta)$ is a contraction for all $n \geq n_0$. This shows that there is a unique fixed point of (3.3.1) in the ball $\mathcal{U}(p_n^\theta u^*, \delta)$.

To show that $u_n^* \xrightarrow{\mathcal{P}^\theta} u^*$, we proceed as follows:

$$\left\| p_n^\theta u^* - u_n^* \right\|_{E_n^\theta} = \left\| \Phi_n(u_n^*) - p_n u^* \right\|_{E_n^\theta}$$

$$\leq \left\| \Phi_n(u_n^*) - \Phi_n(p_n u^*) \right\|_{E_n^\theta} + \left\| \Phi_n(p_n u^*) - p_n u^* \right\|_{E_n^\theta}.$$

The fact that Φ_n is a uniform contraction in E_n^θ implies that

$$\left\| p_n^\theta u^* - u_n^* \right\|_{E_n^\theta} \leq 2 \left\| \Phi_n(p_n^\theta u^*) - p_n^\theta u^*) \right\|_{E_n^\theta}.$$

From (3.3.2) we conclude the required convergence. □

As an immediate consequence of this proposition, we obtain the following theorem.

Theorem 3.12. *Assume that Eq. (3.1.1) has exactly m solutions $u^1, \ldots, u^m$ and all of them are hyperbolic. Assume that for any point u^i and any $\epsilon > 0$, there exists $\delta > 0$ such that*

$$\left\| f_n'(u_n) - f_n'(p_n^\theta u^i) \right\|_{E_n} \leq \epsilon \quad \text{if} \quad \left\| u_n - p_n^\theta u^i \right\|_{E_n^\theta} \leq \delta.$$

Then there exists n_0 such that for all $n \geq n_0$, the problems (3.3.1) have exactly m solutions $u_n^1, \ldots, u_n^m$. Moreover, we have

$$\left\| u_n^k - p_n^\theta u^k \right\|_{E_n^\theta} \to 0 \quad \text{as } n \to \infty \text{ for any } k = 1, \ldots, m.$$

Proof. Due to Proposition 3.3, we have that any solution u_n^k of (3.3.1), $n \to \infty$, lies in a neighborhood of the set of equilibria of Eq. (3.1.1). Proposition 3.5 implies that a neighborhood of u^k contains only one solution of (3.3.1), which converges to u^k. This proves the result. □

3.4 Approximation of Unstable Manifolds

Recall that we assume that all equilibrium solutions of the problem (1.2.16) are hyperbolic. We also assume that each solution u^* of Eq. (3.1.1) is such that the spectrum $\sigma(A + f'(u^*))$ of the operator $A + f'(u^*)$ does not intersect with the imaginary axis.

Proposition 3.6. *Let A be a sectorial operator with compact resolvent and $f'(u^*) \in B(E^\theta, E)$. Then the operator $(A + f'(u^*))$ is sectorial with compact resolvent.*

This assertion immediately follows from [129, Theorems 1.3.2 and 1.4.8]. From Proposition 3.6,

$$\sigma^+ = \sigma\big((A + f'(u^*))\big) \cap \{\lambda \in \mathbb{C} : \operatorname{Re}\lambda > 0\}$$

consists of a finite number (possibly zero) of eigenvalues with finite multiplicities and there is $\beta > 0$ and a finite rank projector $P(\sigma^+)$ such that

$$\begin{cases} \left\|e^{(A+f'(u^*))t}P\right\|_{B(E)} \le Me^{\beta t}, & t \le 0, \\ \left\|e^{(A+f'(u^*))t}(I - P)\right\|_{B(E,E^\theta)} \le Mt^{-\theta}e^{-\beta t}, & t \ge 0,\ \theta \ge 0. \end{cases} \tag{3.4.1}$$

Proposition 3.7. *Let $u_n^* \xrightarrow{P^\theta} u^*$, where u_n^* is a solution of (3.3.1) and u^* is a hyperbolic solution of Eq. (3.1.1). Assume that*

$$(\lambda I_n - A_n)^{-1} \xrightarrow{PP} (\lambda I - A)^{-1} \ \text{compactly.}$$

Then there is a number $n \ge n_0$ such that $\sigma(A_n + f_n'(u_n^*))$ does not intersect the imaginary axis for $n \ge n_0$ and

$$\left\|(A_n + f_n'(u^*))^{-1}\right\| \le C,$$

where C is independent of n. Moreover, if P_n is the projector defined by the spectral set

$$\sigma_n^+ = \big\{\lambda \in \sigma(A_n + f_n'(u_n^*)) : \ \operatorname{Re}\lambda > 0\big\},$$

then P_n converges compactly to P as $n \to \infty$ (hence $\operatorname{rank}(P_n) = \operatorname{rank}(P)$ for $n \ge n_0$) and the family of sets σ_n^+ converges, $\sigma_n^+ \to \sigma^+$, in the sense of (i)–(iii) of Theorem 3.5 and (iv)′–(v) of Theorem 3.6.

Proposition 3.7 follows from Lemma 3.8 and Theorem 3.6.

Proposition 3.8. *Assume that*

$$(\lambda I_n - A_n)^{-1} \xrightarrow{PP} (\lambda I - A)^{-1} \ \text{compactly.}$$

Then for sufficiently large n, the equilibrium u_n^* of (3.3.1) is hyperbolic and there is $\beta > 0$ such that if P_n is the projector determined by σ_n^+, then

$$\begin{cases} \left\|e^{t(A_n+f_n'(u_n^*))}P_n\right\|_{B(E_n)} \le Me^{\beta t}, & t \le 0, \\ \left\|e^{t(A_n+f_n'(u^*))}(I_n - P_n)\right\|_{B(E_n,E_n^\theta)} \le Mt^{-\theta}e^{-\beta t}, & t \ge 0,\ \theta \ge 0. \end{cases} \tag{3.4.2}$$

Proposition 3.8 is a direct consequence of Theorem 3.17.

Theorem 3.13. *Let* $\Delta_{cc} \neq \varnothing$ *and* u^* *be a hyperbolic equilibrium point of the problem* (1.2.16). *Then for sufficiently small* $\delta > 0$, *Eq.* (3.1.11) *has a solution* $\zeta^*(\cdot)$ *such that the equations*

$$\zeta_n(t) = \Theta_n(\eta_n, \zeta_n(t))$$

have solutions $\zeta_n^*(\cdot)$, $n \geq n_0$, *and* $\zeta_n^*(t) \to \zeta^*(t)$ *uniformly in* $t \in (-\infty, 0]$ *as* $n \to \infty$.

Proof. We know that $\left| \mathrm{ind}(\zeta_0, I - \Theta) \right| = 1$ for $\zeta_0 = 0$. According to [168, Theorem 54.1] there exist ρ and $\delta > 0$ such that $\left| \mathrm{ind}(\zeta, I - \Theta) \right| = 1$ for $\|\eta - 0\| \leq \delta$ and $\|\zeta - 0\| \leq \rho$. By Theorem 3.10, we see that $\Theta_n \to \Theta$ compactly, so

$$\gamma(I_n - \Theta_n, \partial\Omega_n^{M,\omega}) = \mathrm{ind}(\zeta, I - \Theta)$$

and, therefore, the equation $\zeta_n(t) = \Theta_n(\eta_n, \zeta_n(t))$ has at least one solution ζ_n^* in any set $\Omega_n^{M,\omega}$, $n \geq n_0$. The sequence $\{\zeta_n^*\}$ is discretely compact and $\zeta_n^* \to \zeta^*$ as $n \to \infty$. $\qquad\square$

Proposition 3.9. *Assume also that* u^* *is a hyperbolic equilibrium for the problem* (1.2.16) *and zero does not belong to the spectrum of the operator*

$$A + f'(u^*) : D(A) \subset E \to E.$$

Assume that

$$\left(\lambda I_n - A_n\right)^{-1} \xrightarrow{\mathcal{PP}} (\lambda I - A)^{-1} \quad \textit{compactly},$$
$$\left\| f_n(u_n) - f_n'(p_n^\theta u^*) \right\| \to 0 \quad \textit{uniformly in } n$$

as $\|u_n - p_n^\theta u^*\|_{E_n^\theta} \to 0$. *Then there exist* δ *and* $n_0 > 0$ *such that* u_n^* *has a local unstable manifold* $W_{loc}^u(u_n^*) \subset E_n^\theta$ *for* $n \geq n_0$. *If we introduce the notation*

$$W_{n,\delta}^u(u_n^*) = \left\{ w_n \in W_{loc}^u(u_n^*) : \left\| w_n - u_n^* \right\|_{E_n^\theta} < \delta \right\}, \quad n \geq n_0,$$

then $W_{n,\delta}^u(u_n^*)$ *converges to* $W_\delta^u(u^*)$ *as* $n \to \infty$, *i.e.,*

$$\sup_{w_n \in W_{n,\delta}^u(u_n^*)} \inf_{w \in W_\delta^u(u^*)} \left\| w_n - p_n^\theta w \right\|_{E_n^\theta}$$
$$+ \sup_{w \in W_\delta^u(u^*)} \inf_{w_n \in W_{n,\delta}^u(u_n^*)} \left\| w_n - p_n^\theta w \right\|_{E_n^\theta} \to 0 \quad \textit{as } n \to \infty.$$

Proof. Note that by Proposition 3.5, each of the problems (3.3.1) has a unique solution u_n^* near u^* and, moreover, $\left\| u_n^* - p_n^\theta u^* \right\|_{E_n^\theta} \to 0$ as $n \to \infty$. Due to Proposition 3.7, this implies that the spectrum of $A_n + f_n'(u_n^*)$ is continuous as $n \to \infty$.

Also, for some positive M and β independent of n, according to Proposition 3.8 we have

$$
\begin{aligned}
\left\| e^{\tilde{A}_n t} z_n \right\|_{E_n^\theta} &\leq M e^{-\beta t} \| z_n \|_{E_n^\theta}, & t &\geq 0, \\
\left\| e^{\tilde{A}_n t} z_n \right\|_{E_n^\theta} &\leq M t^{-\theta} e^{-\beta t} \| z_n \|_{E_n}, & t &> 0, \qquad (3.4.3) \\
\left\| e^{B_n t} v_n \right\|_{W_n} &\leq M e^{\beta t} \| v_n \|_{W_n}, & t &\leq 0.
\end{aligned}
$$

Now we will show that for a sufficiently small $\rho > 0$, there is an unstable manifold for u_n^*

$$
W_n^u = \left\{ (v_n(\cdot), z_n(\cdot)) : \ z_n(\cdot) \in (I_n - P_n) E_n^\theta, \ v_n(\cdot) \in W_n \right\}.
$$

As in Theorem 3.1, we introduce the operators Θ_n and the vector $\zeta_n(t) = \begin{pmatrix} v_n(t) \\ z_n(t) \end{pmatrix}$. From the inequalities (3.2.2) we obtain that there is positive δ such that the operators $\Theta_n(\eta_n, \cdot)$ are Fréchet differentiable in the balls $\mathcal{U}(p_n^\theta \zeta^*, \delta)$ and for any $\epsilon > 0$ there is $\delta_\epsilon > 0$ such that

$$
\left\| \Theta_n'(\eta_n, \zeta_n) - \Theta_n'(\eta_n, p_n^\theta \zeta^*) \right\| \leq \epsilon \quad \text{for} \quad \left\| \zeta_n(\cdot) - p_n^\theta \zeta^*(\cdot) \right\|_{\Upsilon_n} \leq \delta_\epsilon.
$$

It is also clear that the conditions

$$
\left\| p_n^\theta \zeta^*(\cdot) - \Theta_n(\eta_n, p_n^\theta \zeta^*(\cdot)) \right\|_{\Upsilon_n} \to 0,
$$

$$
\left\| \left(I_n - \Theta_n'(\eta_n, p_n^\theta \zeta^*(\cdot)) \right)^{-1} \right\| \leq \text{const}
$$

as $n \to \infty$ are satisfied. Then by [288, Theorem 2] we conclude that there exist $n_0 \in \mathbb{N}$ and $0 < \delta_0 \leq \delta$ such that each of the problems (3.2.3) has a unique solution $\zeta_n^*(\cdot)$ in the ball $\mathcal{U}(p_n^\theta \zeta^*, \delta_0)$ for $n \geq n_0$ and $\zeta_n^* \to \zeta^*$ as $n \to \infty$. $\qquad \square$

The following assertion is an immediate consequence of this proposition.

Corollary 3.2. *Assume that the conditions of Proposition 3.9 hold, Eq. (3.1.1) has exactly m solutions $u^1, \ldots, u^m$, and all of them are hyperbolic. Then there exists sufficiently small $\delta > 0$ such that each of the problems (3.3.1) has exactly m solutions and their local unstable manifolds $W_\delta^u(u_n^k)$, $k = 1, \ldots, m$, behave continuously in E_n^θ as $n \to \infty$.*

In this way, one gets lower semicontinuity of attractors.

Theorem 3.14. *Let $\mathcal{A}$ and $\mathcal{A}_n$ be attractors for the problems (1.2.16) and (1.4.5), respectively. Assume that the problem (1.2.16) has gradient structure and that all its equilibrium points are hyperbolic. If $x^* \in \mathcal{A}$, then there exists a sequence of points $\{x_n^*\}$, $x_n^* \in \mathcal{A}_n$, such that $x_n^* \xrightarrow{\mathcal{P}^\theta} x^*$ and the sequence of attractors $\{\mathcal{A}_n\}$ is $\mathcal{P}^\theta$-lower semicontinuous at infinity.*

Proof. Assume that $x^* \in \mathcal{E}$. By Proposition 3.4, the statement of theorem follows. Now assume that x^* lies on the unstable manifold. Then, moving back in time by some trajectory from the point x^*, one can come to a point y from a neighborhood of some equilibrium point, which—as we know from Theorem 3.13—is approximated by points $y_n \in \mathcal{A}_n$. Starting now from the points y_n, according to Theorem 2.3, we obtain the statement of Theorem 3.14. $\qquad\square$

3.5 Auxiliary Results

Lemma 3.3. *Assume that $B_n \xrightarrow{\mathcal{PP}} B$ compactly, the operators B and B_n are compact, and $\mathcal{N}(I + B) = \{0\}$. Then there exist an $n_0 > 0$ and $M > 0$ such that*

$$\left\|(I_n + B_n)^{-1}\right\|_{B(E_n)} \leq M, \quad n \geq n_0. \tag{3.5.1}$$

Proof. From the compact convergence $B_n \xrightarrow{\mathcal{PP}} B$, we deduce the regular convergence $I_n + B_n \xrightarrow{\mathcal{PP}} I + B$; hence the result follows from Theorem 3.4. $\qquad\square$

Lemma 3.4. *Let $\Delta_{cc} \neq \varnothing$. Then for any $\lambda \in \rho(A)$, there is a $n_\lambda > 0$ such that $\lambda \in \rho(A_n)$ for all $n \geq n_\lambda$ and there is a constant $M_\lambda > 0$ such that*

$$\left\|(\lambda I_n - A_n)^{-1}\right\| \leq M_\lambda, \quad n \geq n_\lambda.$$

Moreover, $(\lambda I_n - A_n)^{-1} \xrightarrow{\mathcal{PP}} (\lambda I - A)^{-1}$ compactly.

Proof. The result follows from Lemma 3.3 and Corollary 3.1. $\qquad\square$

Lemma 3.5. *Let $\Delta_{cc} \neq \varnothing$. Assume that the resolvents $(\lambda I - A)^{-1}$ and $(\lambda I_n - A_n)^{-1}$ are compact. Let Λ be a compact subset of $\rho(A)$. Then there is a constant $n_\Lambda > 0$ such that $\Lambda \subset \rho(A_n)$ for all $n \geq n_\Lambda$ and*

$$\sup_{\substack{\lambda \in \Lambda, \\ n \geq n_\Lambda}} \left\|(\lambda I_n - A_n)^{-1}\right\| < \infty. \tag{3.5.2}$$

Moreover, for any $u \in E$

$$\sup_{\lambda \in \Lambda} \left\| \left(\lambda I_n - A_n\right)^{-1} p_n u - p_n \left(\lambda I - A\right)^{-1} u \right\| \to 0. \qquad (3.5.3)$$

Proof. First, we prove that there is a number $n_\Lambda > 0$ such that $\Lambda \subset \rho(A_n)$ for all $n \geq n_\Lambda$. Assume the contrary; then there are sequences $k_n \to \infty$ and $\{\lambda_{k_n}\} \in \Lambda$ such that λ_{k_n} is an eigenvalue of A_{k_n}. Since Λ is compact, we may assume that there is $\bar{\lambda} \in \Lambda$ such that $\lambda_{k_n} \to \bar{\lambda}$. It follows from Theorem 3.5(ii) that $\bar{\lambda} \in \sigma(A)$, a contradiction.

To prove (3.5.2), it suffices to prove that

$$\sup_{\substack{\lambda \in \Lambda, \\ n \geq n_\Lambda}} \left\| \left(I - \lambda A_n^{-1}\right)^{-1} \right\| < \infty.$$

Assume the contrary, i.e., let $k_n \to \infty$ and $\lambda_{k_n} \in \Lambda$ be sequences (we may assume that λ_{k_n} converges to $\bar{\lambda} \in \Lambda$) such that

$$\left\| \left(I - \lambda_{k_n} A_{k_n}^{-1}\right)^{-1} \right\| \to \infty.$$

Since $\lambda_{k_n} A_{h_n}^{-1}$ converges compactly to $\bar{\lambda} A^{-1}$, this contradicts Lemma 3.3.

It remains to prove (3.5.3). Once again, we prove it by contradiction. Assume that there are sequences $k_n \to \infty$ and $\Lambda \ni \lambda_{k_n} \to \bar{\lambda} \in \Lambda$ and $\epsilon > 0$ such that

$$\left\| \left(\lambda_{k_n} I_{k_n} - A_{k_n}\right)^{-1} p_{k_n} u - p_{k_n} \left(\lambda_{k_n} I - A\right)^{-1} u \right\| \geq \epsilon. \qquad (3.5.4)$$

This contradicts the fact

$$\left\| \left(\lambda I_n - A_n\right)^{-1} p_n u - p_n \left(\lambda I - A\right)^{-1} u \right\| \to 0 \quad x\text{for } \lambda = \bar{\lambda}$$

and the estimate (3.5.2). $\qquad \square$

Next we show that the compact convergence of A_n^{-1} to A^{-1} ensures that the resolvent set of A_n contains a fixed sector for almost all n. Introduce the notation

$$\Sigma_{\phi, \bar{\omega}} = \left\{ \lambda \in \mathbb{C} : \left| \arg(\lambda - \bar{\omega}) \right| < \pi - \phi \right\}.$$

Lemma 3.6. *Assume that $\Delta_{cc} \neq \varnothing$ and Conditions* (1.1.3) *and* (B$_1$) *are fulfilled. Then there are a sector $\Sigma_{\bar{\phi}, -\omega}$, $\phi < \bar{\phi} < \pi/2$, $0 < \omega < \tilde{\omega}$, a constant $n_\omega > 0$, and a constant M_ω such that*

$$\left\| \left(\lambda I_n - A_n\right)^{-1} \right\| \leq \frac{M_\omega}{|\lambda - \omega|} \quad \text{for all } \lambda \in \Sigma_{\bar{\phi}, -\omega} \text{ and } n \geq n_\omega. \qquad (3.5.5)$$

Proof. Let $\omega < \omega_1 < \tilde{\omega}$. Consider the domain

$$\Lambda = \left\{ \lambda \in \mathbb{C} : \ \operatorname{Re} \lambda \geq \omega_1, \lambda \notin \Sigma_{\phi,\tilde{\omega}} \right\}.$$

Then Λ is a compact set and $\Lambda \subset \rho(A)$. Lemma 3.5 implies that there is a constant $n_\Lambda > 0$ such that $\Lambda \subset \rho(A_n)$ for all $n \geq n_\Lambda$ and

$$\sup_{\substack{\lambda \in \Lambda, \\ n \geq n_\Lambda}} \left\| (\lambda I_n - A_n)^{-1} \right\| < \infty. \tag{3.5.6}$$

If we take $\bar{\phi} < \pi/2$ such that

$$\Sigma_{\bar{\phi},\omega} \cap \left\{ \lambda \in \Sigma_{\phi,\tilde{\omega}} : \ \operatorname{Re} \lambda < \omega_1 \right\} = \varnothing,$$

we have that $\Sigma_{\bar{\phi},\omega} \subset \rho(A_n)$ for all $n \geq n_\Lambda$. The estimate (3.5.5) follows easily from (3.5.6) and the estimate (B_1). $\qquad\square$

Lemma 3.7. *Let A_n and A be such that the conditions* (1.1.3), (B_1), *and* $\Delta_{cc} \neq \varnothing$ *are satisfied. Then*

$$A_n^{-\theta} \xrightarrow{\mathcal{PP}} A^{-\theta} \quad compactly$$

for any $0 < \theta < 1$.

Proof. Let $\Sigma_{\bar{\phi},-\omega}$ be as in Lemma 3.6 and Ω be the boundary of $\Sigma_{\bar{\phi},-\omega}$. We have

$$(-A_n)^{-\theta} = \frac{1}{2\pi i} \int_\Omega (-\lambda)^{-\theta} (\lambda I_n - A_n)^{-1} d\lambda.$$

By Lemma 3.6, the above integral absolutely converges uniformly for $n \geq n_\omega$. Therefore, given $\epsilon > 0$, one can partition the contour $\Omega = \Omega_1^\epsilon \cup \Omega_2^\epsilon$ such that Ω_1^ϵ is bounded and

$$\frac{1}{2\pi i} \int_{\Omega_2^\epsilon} \left\| (-\lambda)^{-\theta} (\lambda I_n - A_n)^{-1} \right\| d\lambda \leq \epsilon, \quad n \geq n_\omega.$$

We rewrite the integral over Ω_1^ϵ as follows:

$$B_n := \frac{A_n^{-1}}{2\pi i} \int_{\Omega_1^\epsilon} (-\lambda)^{-\theta} \lambda (\lambda I_n - A_n)^{-1} d\lambda$$

and observe that $\left\| (-\lambda)^{-\theta} \lambda (\lambda I_n - A_n)^{-1} \right\|$ is bounded uniformly for $n \geq n_\omega$. The fact that $A_n^{-1} \to A^{-1}$ compactly implies that $B_n \to B$ compactly. Let μ be a noncompactness measure. Now, taking arbitrary sequences $n \to \infty$ and $\{u_n\}$, $u_n \in E_n$, $\|u_n\| = 1$, we obtain that

$$\mu\left(\left\{ (-A_n)^{-\theta} u_n \right\} \right) \leq \mu\left(B_n \{u_n\} \right)$$

$$+ \mu\left(\frac{1}{2\pi i} \int_{\Omega_2^\epsilon} (-\lambda)^{-\theta} (\lambda I_n - A_n)^{-1} d\lambda \, u_n \right) \leq \epsilon.$$

It follows that for given sequences $n \to \infty$ and $\{u_n\}$ such that $u_n \in E_n$, $\|u_n\| = 1$, we obtain that $\mu(\{(-A_n)^{-\theta} u_n\}) = 0$ and $(-A_n)^{-\theta} \to (-A)^{-\theta}$ compactly.

To see that $(-A_n)^{-\theta} \xrightarrow{\ \mathcal{PP}\ } (-A)^{-\theta}$, we use the fact that

$$(\lambda I_n - A_n)^{-1} \xrightarrow{\ \mathcal{PP}\ } (\lambda I - A)^{-1} \quad \text{for all } \lambda \in \Omega \text{ as } n \to \infty$$

and apply the dominated convergence theorem.   □

Now we consider relatively bounded perturbations of A_n and A in the following way. Assume that perturbations satisfy the following condition:

(L) $D_n \in B(E_n^\theta, E_n)$ and $D \in B(E^\theta, E))$ such that D_n and D are consistent.

We also impose the following hyperbolicity condition:

(H) $\sigma(A + D) \cap \{\lambda \in \mathbb{C} : \operatorname{Re}\lambda = 0\} = \varnothing$.

It is clear that if the resolvent of A is compact, then the operator $A + D$ has a compact resolvent.

Lemma 3.8. *Assume that $\Delta_{cc} \neq \varnothing$ and the conditions (1.1.3) and (B$_1$) are fulfilled. If for any $0 \leq \xi < 1$, the operators $D_n \in B(E_n^\theta, E_n)$ are such that $D_n \to D$ in the sense of the spaces E_n^θ, E_n and E^θ, E, then*

$$(-A_n)^\xi (A_n + D_n)^{-1} \xrightarrow{\ \mathcal{PP}\ } (-A)^\xi (A + D)^{-1} \quad \text{compactly.}$$

Proof. We choose positive numbers r_1 and r_2 such that $1 > \theta + r_1 \geq \xi$ and $r_2 = 1 - \theta - r_1$ and note that

$$(-A_n)^\xi (A_n + D_n)^{-1}$$
$$= (-A_n)^{-(\theta + r_1 - \xi)} \Big(-I_n + (-A_n)^{-r_2} D_n (-A_n)^{-\theta - r_1} \Big)^{-1} (-A_n)^{-r_2}.$$

Since $\{(-A_n)^{-r_2} D_n (-A_n)^{-\theta - r_1}\}$ converges compactly, the result now follows from Lemmas 3.7 and 3.3.   □

Lemma 3.9. *Let A_n and A be such that the conditions (1.1.3), (B$_1$), $\Delta_{cc} \neq \varnothing$, (L), and (H) are fulfilled. Then there are a sector $\Sigma_{\phi_d, \omega_d}$, $\phi_d < \pi/2$, $\omega_d \in \mathbb{R}$, a constant $n_d > 0$, and a constant M_d such that*

$$\left\| \left(\lambda I_n - (A_n + f_n'(u_n^*)) \right)^{-1} \right\| \leq \frac{M_d}{|\lambda - \omega_d|} \qquad (3.5.7)$$

for all $\lambda \in \Sigma_{\phi_d, \omega_d}$ and all $n \geq n_d$.

Proof. Lemma 3.6 and the moment inequality imply that

$$\left\|(-A_n)^\theta(\lambda I_n - A_n)^{-1}\right\| \le \frac{M}{|\lambda|^{1-\theta}}, \quad n \ge n_\omega,$$

for $\lambda \in \Sigma_{\bar{\phi},0}$. Since $D_n(-A_n)^{-\theta}$ is bounded uniformly in n and

$$\left(\lambda I_n - A_n - D_n\right) = \left(I_n - D_n(-A_n)^{-\theta}(-A_n)^\theta(\lambda I_n - A_n)^{-1}\right)\left(\lambda I_n - A_n\right),$$

we see that there is $R > 0$ such that for $|\lambda| > R$, $\lambda \in \Sigma_{\bar{\phi},0}$, the operator

$$I_n - D_n(-A_n)^{-\theta}(-A_n)^\theta(\lambda I_n - A_n)$$

is invertible and

$$\left\|\left(I_n - D_n(-A_n)^{-\theta}(-A_n)^\theta\left(\lambda I_n - A_n\right)^{-1}\right)^{-1}\right\| \le 2.$$

This implies the required result (see Fig. 3.3) for any $\omega_d > R$ with appropriately chosen θ_d such that $\Sigma_{\theta_d,\omega_d} \subset \left\{\lambda \in \Sigma_{\bar{\theta},0} : |\lambda| > R\right\}$. $\square$

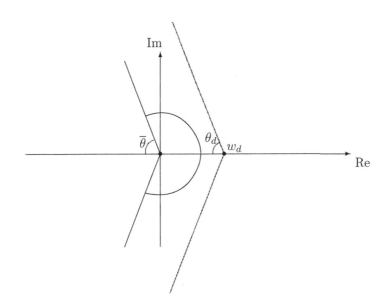

Fig. 3.3

Observing that, due to Lemma 3.8, $(A_n + f_n'(u_n^*))^{-1}$ converges compactly to $(A + f'(u^*))^{-1}$ and proceeding exactly as in the proof of Lemma 3.5, we obtain the following result.

Lemma 3.10. *Let A_n and A be such that the conditions (1.1.3), (B$_1$), $\Delta_{cc} \neq \varnothing$, (L), and (H) are fulfilled and let Λ be a compact subset of $\rho((A + f'(u^*)))$. Then there is a constant $n_\Lambda > 0$ such that*

$$\Lambda \subset \rho\big((A_n + f'_n(u_n^*))\big) \quad \text{for all } n \geq n_\Lambda$$

and

$$\sup_{\substack{\lambda \in \Lambda, \\ n \geq n_\Lambda}} \left\| \left(\lambda I_n - \big(A_n + f'_n(u_n^*) \big) \right)^{-1} \right\| < \infty. \tag{3.5.8}$$

Moreover,

$$\sup_{\lambda \in \Lambda} \left\| \left(\lambda I_n - \big(A_n + f'_n(u_n^*) \big) \right)^{-1} p_n u - \left(\lambda I - \big(A + f'(u^*) \big) \right)^{-1} u \right\| \to 0 \tag{3.5.9}$$

for any $u \in E$.

To an isolated point $\lambda \in \sigma((A + f'(u^*)))$, we associate its generalized eigenspace $W(\lambda, (A + f'(u^*))) = P(\lambda, (A + f'(u^*)))E$, where

$$P\big(\lambda, (A + f'(u^*))\big) = \frac{1}{2\pi i} \int_{|\xi - \lambda| = \delta} \left(\xi I - \big(A + f'(u^*) \big) \right)^{-1} d\xi$$

and δ is chosen so small that there is no point of $\sigma\big((A + f'(u^*))\big)$ in the disc $\big\{ \xi \in \mathbb{C} : |\xi - \lambda| \leq \delta \big\}$. We introduce the notation

$$W_n\Big(\lambda, \big(A_n + f'_n(u_n^*) \big)\Big) = P_n\Big(\lambda, \big(A_n + f'_n(u_n^*) \big)\Big) E_n,$$

where

$$P_n\Big(\lambda, \big(A_n + f'_n(u_n^*) \big)\Big) = \frac{1}{2\pi i} \int_{|\xi - \lambda| = \delta} \left(\xi I_n - \big(A_n + f'_n(u_n^*) \big) \right)^{-1} d\xi.$$

Theorem 3.15. *Let A_n and A be such that the conditions (1.1.3), (B$_1$), $\Delta_{cc} \neq \varnothing$, (L), and (H) are fulfilled. Then the following statements hold:*

(i) *there exists $n_0 > 0$ such that*

$$\dim W_n\big(\sigma_n^+, (A_n + f'_n(u_n^*))\big) = \dim W\big(\sigma^+, (A + f'(u^*))\big)$$

for all $n \geq n_0$;

(ii) *for any $u \in W\big(\sigma^+, (A + f'(u^*))\big)$, there exists a sequence $\{u_n\}$, $u_n \in W_n\big(\sigma^+, (A_n + f'_n(u_n^*))\big)$, such that $u_n \xrightarrow{\mathcal{P}^\theta} u$;*

(iii) *any sequence $\{u_n\}$, $u_n \in W_n\big(\sigma_n^+, (A_n + f'_n(u_n^*))\big)$, $n \in \mathbb{N}$, such that $\|u_n\|_{X_n} = 1$ has a convergent subsequence and any limit point of this sequence belongs to $W\big(\sigma^+, (A + f'(u^*))\big)$;*

(iv) *there exist $\beta > 0$ and $n_\beta > 0$ such that*

$$\sigma\big((A_n + f_n'(u_n^*))\big) \cap \{\lambda \in \mathbb{C} : \ |\operatorname{Re}\lambda| > \beta\} = \varnothing, \quad n \geq n_\beta;$$

(v) *there exist ϕ_β $(0 < \phi_\beta < \pi/2)$, $\omega_\beta > 0$, and $n_\beta > 0$ such that*

$$\Sigma_{\phi_\beta, -\omega_\beta} \subset \rho(A_n), \quad n \geq n_\beta,$$

$$\left\|\big(\lambda I_n - (A_n + f_n'(u_n^*))\big)^{-1}\right\| \leq \frac{M_\beta}{|\lambda - \omega_\beta|} \tag{3.5.10}$$

for all $\lambda \in \Sigma_{\phi_\beta, -\omega_\beta}$ and all $n \geq n_\beta$.

3.5.1 Convergence of linear semigroups and uniform estimates

It is well known that if the condition (1.1.3) is fulfilled, then for each $\omega < \tilde{\omega}$, there is a constant $M_\omega \geq 1$ such that

$$\left\|t^\theta e^{At}\right\|_{B(E, E^\theta)} \leq M_\omega e^{-\omega t} \quad \forall t \geq 0.$$

Lemma 3.6 implies that an estimate similar to (1.1.3) holds uniformly for the semigroups generated by A_n for sufficiently large n. Hence we obtain the following result.

Theorem 3.16. *Assume that $\Delta_{cc} \neq \varnothing$ and the conditions (1.1.3) and (B_1) are fulfilled. Then for each $\omega < \tilde{\omega}$, there exist constants n_0 and M_ω such that*

$$\left\|t^\theta e^{A_n t}\right\|_{B(E_n, E_n^\theta)} \leq M_\omega e^{-\omega t} \quad \forall t \geq 0, \quad 0 \leq \theta \leq 1, \quad n \geq n_0. \tag{3.5.11}$$

Proof. Consider the sector

$$\Sigma(\pi - \phi, \omega) = \left\{\lambda \in \mathbb{C} : \ \big|\arg(\lambda - \omega)\big| < \pi - \phi\right\}, \quad \omega \in \mathbb{R}, \quad 0 < \phi < \frac{\pi}{2}.$$

From Condition (B_1) we obtain suitable numbers $\omega_3 \in \mathbb{R}$, $n_\omega \in \mathbb{N}$, and $0 < \phi < \pi/2$ such that

$$\left\|\big(\lambda I_n - A_n\big)^{-1}\right\| \leq \frac{M_3}{|\lambda - \omega_3|} \tag{3.5.12}$$

for all $\lambda \in \Sigma(\pi - \phi, \omega_3)$ and $n \geq n_\omega$. From (1.1.3) it follows that $\operatorname{Re}\sigma(A) < \omega'$ for some $\omega' < 0$ and, moreover,

$$\left\|\big(\lambda I - A\big)^{-1}\right\| \leq \frac{M}{|\lambda - \omega'|}$$

for $\operatorname{Re}\lambda > \omega'$. $\qquad\square$

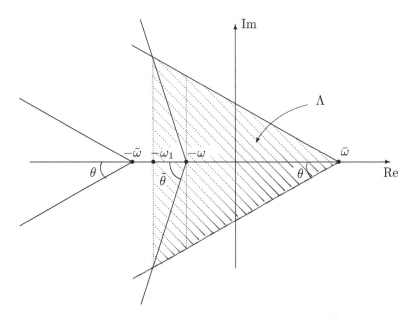

Fig. 3.4

Now let $0 < \tilde{\omega} < |\omega'|$; we consider the triangular domain (see Fig. 3.4)

$$\Lambda = \Big\{ \lambda \in \mathbb{C} : \ \operatorname{Re} \lambda \geq -\tilde{\omega}, \ \lambda \notin \Sigma(\pi - \phi, \omega_3) \Big\}.$$

The following assertion is a consequence of Theorem 3.15.

Theorem 3.17. *Assume that the condition* $\Delta_{cc} \neq \varnothing$ *and the conditions* (1.1.3), (B$_1$), (L), *and* (H) *are satisfied. Then for each* $\omega_\beta < \beta$, *there are constants* n_{ω_β} *and* M_{ω_β} *such that for* $n \geq n_{\omega_\beta}$ *the following inequalities hold:*

$$\Big\| t^\theta e^{(A_n + f'_n(u_n^*))t} \Big(I_n - P_n\big(\sigma_n^+, \big(A_n + f'_n(u_n^*)\big)\big) \Big) \Big\|_{B(E_n, E_n^\theta)} \leq M_{\omega_\beta} e^{-\omega_\beta t}$$

$$\forall t \geq 0, \quad 0 \leq \theta \leq 1,$$

$$\Big\| e^{(A_n + f'_n(u_n^*))t} P_n\big(\sigma_n^+, \big(A_n + f'_n(u_n^*)\big)\big) \Big\|_{B(E_n)} \leq M_{\omega_\beta} e^{\omega_\beta t} \quad \forall t \leq 0.$$

Remark 3.2. The discrete convergence of resolvents (without the condition $\Delta_{cc} \neq \varnothing$) is not sufficient to obtain the statement of Theorem 3.17, which is clear from the following example.

For each $n \geq 1$, let L_n be an $(n \times n)$-matrix all of whose entries equal to zero except for the first upper diagonal which is formed by ones. Let $M_n = -\omega I_n + L_n$. The unique element of the spectrum of M_n is the number $-\omega$. Moreover, since

$$\left\| \exp(t M_n) \right\| \leq e^{-\omega t} e^{\|L_n\| t} \leq e^{(1-\omega)t},$$

is clear that, on finite intervals, we have the strong convergence to $e^{-\alpha t} I$ of the semigroups generated by

$$A_n = \operatorname{diag}\left(-\alpha I_1, -\alpha I_2, \ldots, M_n, 0, 0 \ldots \right).$$

On the other hand, the $(1, n)$-element of $\exp(t M_n)$ is nothing but

$$e^{-\omega t} \frac{t^{n-1}}{(n-1)!},$$

so that, by the Stirling formula,

$$\left\| e^{(n-1) M_n} \right\| \geq e^{-\omega(n-1)} \frac{(n-1)^{n-1}}{(n-1)!} \equiv \frac{e^{(n-1)(1-\omega)}}{(2\pi n)^{1/2}},$$

and if $0 < \omega < 1$, we have

$$\sup\left\{ \left\| e^{(n-1) M_n} \right\| : n \geq 1, \ t \geq 0 \right\} = +\infty.$$

Chapter 4

Shadowing
for Abstract Parabolic Equations

Classical shadowing results state that pseudo-trajectories of a finite-dimensional dynamical system can be "shadowed" by true trajectories provided that the system possesses a certain kind of hyperbolicity. This is usually stated in a more quantitative form as follows. For any given $\varepsilon > 0$, there exists some $\delta > 0$ such that for any pseudo-trajectory that allows jumps of magnitude δ at successive time instances, there exists a true trajectory within an ε-neighborhood of the pseudo-trajectory uniformly in time. Such statements hold for both time-continuous and time-discrete dynamical systems in neighborhoods of (suitably defined) hyperbolic sets. We refer the reader to [105, 222, 227] for an excellent summaries of various shadowing results.

If numerical approximations must be included into such an approach, it becomes clear that the concept of a pseudo-trajectory needs a considerable extension for a shadowing principle to be still valid. For example, the discretization of an autonomous ODE by a one-step method leads to a mapping depending on step-size, and shadowing now means approximation of a time-discrete orbit by a true continuous trajectory or vice versa. A result of this type was derived in [52] near equilibrium hyperbolic points and for more general hyperbolic situations in [104].

Shadowing results for numerical approximations of time-dependent partial differential equations usually involve both time and space discretization, i.e., continuous trajectory in an infinite-dimensional space should be shadowed by a discrete time trajectory in a finite-dimensional space and vice versa.

Such a result was derived in [176] for a finite-element method combined with the backward Euler discretization in time when applied to a nonlinear reaction diffusion system in a neighborhood of a equilibrium hyperbolic solution. These results extended earlier work on semidiscretizations with finite elements of the same authors [175]. Shadowing results for semidiscretizations in time were obtained earlier in [7] near hyperbolic equilibrium states and, more recently, for a linear but nonautonomous setting in [217].

The purpose of this chapter is to study shadowing properties of rather general spatial discretizations of a nonlinear evolution equation (1.2.16), where A is a closed operator that generates an analytic C_0-semigroup $\exp(tA)$ on E. For the discretization in space, we use the theory of discrete approximations developed in [114, 271, 285, 288, 291]. As is known for stationary problems, this theory provides a unified framework for handling such diverse approximations as (conforming and nonconforming) finite-elements methods, finite-difference methods (see [285]), and perturbations of domains (see [21, 272]).

The main results (see Theorems 4.2 and 4.3) of this chapter show that mild solutions near a hyperbolic equilibrium of the system (4.1.5) can be shadowed on arbitrary large time intervals by corresponding mild solutions of the system (4.2.9) and vice versa. Our approach here follows the general idea of constructing shadowing trajectories from boundary-value problems proposed in [7, 52, 175, 176].

4.1 Preliminaries

In a Banach space E, we consider the following inhomogeneous Cauchy problem:

$$u'(t) = Au(t) + g(t), \qquad t \in [0, T],$$
$$u(0) = u_0,$$

$$(4.1.1)$$

where an operator $A \in \mathcal{C}(E)$ generates a C_0-semigroup and $g(\cdot)$ is a function $[0, T] \to E$. The problem (4.1.1) can be considered in various function spaces. The most popular spaces for which the well-posedness can be proved are $C([0, T]; E)$, $C_0^\alpha([0, T]; E)$, and $L^p([0, T]; E)$ (see [30, 295]).

In general, we consider mild or so-called generalized solutions of (4.1.1), i.e., functions

$$u(t) = \exp(tA)u^0 + \int_0^t \exp\big((t-s)A\big)g(s)ds, \quad t \geq 0. \qquad (4.1.2)$$

In this chapter, as in Chapter 3, we consider the semilinear autonomous abstract parabolic problem

$$u'(t) = Au(t) + f(u(t)), \quad t \geq 0,$$
$$u(0) = u^0 \in E,$$

$$(4.1.3)$$

where the function $f(\cdot) : E \to E$ is locally Lipschitzian, bounded, and continuously Fréchet differentiable. It is well known that under these assumptions, a mild solution of the problem (4.1.3) exists on a certain maximal interval. Moreover, writing the solution as $u(t) = S(t)u^0$, we obtain a nonlinear semigroup $S(\cdot)$ on E that satisfies the formula of variation of constants

$$S(t)u^0 = \exp(tA)u^0 + \int_0^t e^{(t-s)A} f(S(s)u^0)ds, \quad t \geq 0. \tag{4.1.4}$$

In the sequel, let $A : D(A) \subseteq E \to E$ be a closed linear operator such that the condition (1.1.3) holds.

For some $0 < \theta \leq 1$, consider the semilinear equation (4.1.3) in the space E^θ:

$$\begin{aligned} u'(t) &= Au(t) + f(u(t)), \quad t \geq 0, \\ u(0) &= u^0 \in E^\theta, \end{aligned} \tag{4.1.5}$$

where $f(\cdot) : E^\theta \subseteq E \to E$ satisfies the following condition:

(F1) for some $\rho > 0$, the function $f(\cdot) : \mathcal{U}_{E^\theta}(u^*; \rho) \mapsto E$ is continuously Fréchet differentiable and for any $\epsilon > 0$, there is $\delta > 0$ such that

$$\left\| f'(w) - f'(z) \right\|_{B(E^\theta, E)} \leq \epsilon$$

for all $w, z \in \mathcal{U}_{E^\theta}(u^*; \rho)$ such that with $\|w - z\|_{E^\theta} \leq \delta$, where u^* is a hyperbolic equilibrium point of the problem (4.1.5).

Here and in what follows, u^* always denotes a hyperbolic equilibrium of Eq. (4.1.3). Similarly to (3.1.7), by the change of variables $v(\cdot) = u(\cdot) - u^*$, the problem (4.1.5) can be reduced to the form

$$v'(t) = A_{u^*}v(t) + F_{u^*}(v(t)), \quad v(0) = v^0, \quad t \geq 0, \tag{4.1.6}$$

where $v^0 = u^0 - u^*$, and

$$A_{u^*} = A + f'(u^*),$$
$$F_{u^*}(w) = f(w + u^*) - f(u^*) - f'(u^*)w \quad \text{for } \|w\|_{E^\theta} \leq \rho.$$

Recall that $F_{u^*}(w) = f(w + u^*) - f(u^*) - f'(u^*)w$ has order $o(\|w\|_{E^\theta})$ and the operator A_{u^*} generates an analytic C_0-semigroup since $f'(u^*) \in B(E^\theta, E)$ (see [235]).

We assume that the part σ^+ of the spectrum of the operator $A + f'(u^*)$, which is located strictly to the right of the imaginary axis, consists of a finite number of eigenvalues with finite multiplicities. This assumption is satisfied, for example, if the resolvent of the operator A is compact. In the case of a hyperbolic point u^*, there is no points of the spectrum of A_{u^*} on $i\mathbb{R}$.

Remark 4.1. The properties of exponential dichotomy were also studied in [110], where integral manifolds attracting arbitrary solutions at an exponential rate were considered.

Let $U(\sigma^+) \subset \{\lambda \in \mathbb{C} : \operatorname{Re} \lambda > 0\}$ be an open, connected neighborhood of σ^+ whose boundary $\partial U(\sigma^+)$ is a closed rectifiable curve. We decompose E^θ using the Riesz projector defined by σ^+ as in (3.1.5). Due to this definition and the analyticity of the C_0-semigroup $\exp(tA_{u^*})$, we have positive constants M_1 and $\beta > 0$ such that (cf. [129])

$$\left\| \exp(tA_{u^*})v \right\|_{E^\theta} \leq \begin{cases} M_1 e^{-\beta t} \|v\|_{E^\theta}, & t \geq 0, \ v \in (I - P(\sigma^+))E^\theta, \\ M_1 e^{\beta t} \|v\|_{E^\theta}, & t \leq 0, \ v \in P(\sigma^+)E^\theta. \end{cases} \tag{4.1.7}$$

Without loss of generality, we can choose a norm in E^θ such that

$$\|v\|_{E^\theta} = \max \left(\left\| P(\sigma+)v \right\|_{E^\theta}, \left\| (I - P(\sigma+))v \right\|_{E^\theta} \right). \tag{4.1.8}$$

If v^0 is close to 0, i.e., $v^0 \in \mathcal{U}_{E^\theta}(0; \rho)$, then the mild solution $v(t; v^0)$ of (4.1.6) can stay in the ball $\mathcal{U}_{E^\theta}(0; \rho)$ for some time. We will recognize such a solution as a solution of a boundary-value problem, where the stable part is prescribed at the beginning and the unstable part at the end. More precisely, for any $v^-, v^+ \in \mathcal{U}_{E^\theta}(0; \rho)$ and any $0 < T \leq \infty$, we consider the boundary-value problem

$$\begin{cases} v'(t) = A_{u^*}v(t) + F_{u^*}(v(t)), & 0 \leq t \leq T, \\ (I - P(\sigma^+))v(0) = (I - P(\sigma^+))v^-, \quad P(\sigma^+)v(T) = P(\sigma^+)v^+. \end{cases} \tag{4.1.9}$$

In the case where $T = \infty$, the second boundary condition vanishes and the differential equation holds on $[0, \infty)$. A mild solution of the problem (4.1.9) satisfies the integral equation

$$v(t) = \exp\left((t - T)A_{u^*}\right)P(\sigma^+)v^+ + \exp(tA_{u^*})\left(I - P(\sigma^+)\right)v^-$$
$$+ \int_0^T \Gamma_T(t, s)F_{u^*}(v(s))ds, \quad 0 \leq t \leq T, \tag{4.1.10}$$

where the Green function $\Gamma_T(\cdot, \cdot)$ is defined by the formula

$$\Gamma_T(t, s) = \begin{cases} \exp\left((t - s)A_{u^*}\right)\left(I - P(\sigma^+)\right), & 0 \leq s \leq t \leq T, \\ \exp\left((t - s)A_{u^*}\right)P(\sigma^+), & 0 \leq t < s \leq T. \end{cases} \tag{4.1.11}$$

Note that (4.1.7) implies

$$\left\| \Gamma_T(t, s)z \right\|_{E^\theta} \leq M_1 \frac{e^{-\beta|t-s|}}{|t - s|^\alpha} \|z\|_E. \tag{4.1.12}$$

Again, in the case $T = \infty$, we set the term involving v^+ in (4.1.10) equal to zero. The existence and uniqueness of solutions of the problem (4.1.10) is established by the following proposition.

Proposition 4.1 (see [56]). *Let A and $f(\cdot)$ satisfy the conditions above, in particular, let Condition (F1) be satisfied. Then there exists $\hat{\rho} > 0$ such that for any $0 < \hat{\rho}_2 \le \hat{\rho}$, there exists $0 < \hat{\rho}_1 \le \hat{\rho}_2$ possessing the following property:* Eq. (4.1.10) *has a unique solution*

$$v(\cdot) = v(u^+, u^-, \cdot) \in C\big([0, T]; \mathcal{U}_{E^\theta}(0; \hat{\rho}_2)\big)$$

for all $v^\pm \in \mathcal{U}_{E^\alpha}(0, \hat{\rho}_1)$ and all $0 < T \le \infty$. If $T = \infty$, then

$$\|v(t)\|_{E^\theta} \to 0 \quad \text{as } t \to \infty.$$

Proof. We apply Lemma 4.1 with $Y = Z = C([0, T]; E^\theta)$ (in the case where $T = \infty$, we consider the space of continuous and bounded functions). Further, we set $y_0 = 0$ and $H(v) = v - G(u^-, u^+; v)$, where the operator $G(v^-, v^+; v)$ is defined by the right-hand side of (4.1.10). First, note that

$$H'(0) = I - G'_v(u^-, u^+; 0) = I,$$

so that we can take $\sigma = 1$ in Lemma 4.1. For any $v, w \in \mathcal{U}_Y(0, \hat{\rho}_2)$, we have by (4.1.12) and Condition (F1)

$$\left\|\Big(G'_v(v^-, v^+; v) - G'_v(v^-, v^+; w)\Big)u\right\|_Z$$

$$\le \sup_{0 \le t \le T} \left\|\int_0^T \Gamma_T(t, s)\Big(F'_{u^*, v}(v(s)) - F'_{u^*, v}(w(s))\Big)u(s)ds\right\|_{E^\theta}$$

$$\le M_1 \sup_{0 \le t \le T} \int_0^T \frac{e^{-\beta|t-s|}}{|t-s|^\theta}\left\|\Big(f'(v(s) + u^*) - f'(w(s) + u^*)\Big)u(s)\right\|_E ds$$

$$\le M_1 \frac{1}{\beta}\|u\|_Y \sup_{v_1, v_2 \in \mathcal{U}_{E^\theta}(0,\hat{\rho}_2)} \|f'(v_1) - f'(v_2)\|_{B(E^\theta; E)}$$

$$\le \frac{1}{2}\|u\|_Y \tag{4.1.13}$$

for sufficiently small $\hat{\rho}_2$. Finally, we choose $\hat{\rho}_1 = \hat{\rho}_2/(4M_1)$ and obtain

$$\|H(0)\|_{E^\theta} \le \sup_{0 \le t \le T}\Big(M_1 e^{-\beta t}\big\|(I - P(\sigma^+))u^-\big\|_{E^\theta}$$

$$+ M_1 e^{\beta(t-T)}\big\|P(\sigma^+)u^+\big\|_{E^\theta}\Big) \le M_1 \cdot 2\hat{\rho}_1 \le \frac{1}{2}\hat{\rho}_2.$$

Now we consider the case $T = \infty$, where we write (4.1.10) as follows:

$$v(t) = \exp(tA_{u^*})(I - P(\sigma^*))v(0) \tag{4.1.14}$$

$$+ \int_0^t \exp\big((t-s)A_{u^*}\big)\big(I - P(\sigma^+)\big)F_{u^*}(v(s))ds \tag{4.1.15}$$

$$+ \int_t^\infty \exp\big((t-s)A_{u^*}\big)P(\sigma^+)F_{u^*}(v(s))ds. \tag{4.1.16}$$

Now, arguing as above with the space $C([0,\infty); E^\theta)$ replaced by

$$C_0([0,\infty); E^\theta) = \Big\{u(\cdot) \in C([0,\infty); E^\theta) : \|u(t)\|_{E^\theta} \to 0 \text{ as } t \to \infty\Big\},$$

we note that $G(u^-, u^+, \cdot)$ maps this space into itself since the fact that $\|v(t)\|_{E^\theta} \to 0$ as $t \to \infty$ implies

$$\big\|F_{u^*}(v(t))\big\|_E \to 0 \quad \text{as } t \to \infty. \tag{4.1.17}$$

The operator $G(v^-; v)$ defined by the right-hand side of (4.1.16) is continuous in both arguments and maps the space $C_0([0;\infty); E^\theta))$ into itself. Indeed, for $t \geq T$ we have

$$\big\|(-A)^\theta G(v^-; v)(t)\big\|_E \leq M_1 e^{-t\beta}\big\|(I - P(\sigma^+))v^-\big\|_{E^\theta}$$

$$+ M_1 e^{-\beta(t-T)} \int_0^T \frac{e^{-\beta(T-s)}}{|t-s|^\alpha}\big\|F_{u^*}(v(s))\big\|_E ds$$

$$+ \int_T^\infty M_1 \frac{e^{-\beta|t-s|}}{|s-t|^\theta}\big\|F_{u^*}(v(s))\big\|_E ds. \tag{4.1.18}$$

Given $\epsilon > 0$, we first take T so large that the first term and the second integral are less than $\epsilon/3$ for all $t \geq T$; then we choose t so large that the first integral is less than $\epsilon/3$.

So, there is a unique solution of the equation $v(\cdot) = G(v^-; v)$ in the space $C_0([0;\infty); E^\theta)$ and the result follows by the uniqueness. $\qquad\square$

In the proof of Proposition 4.1 we used the following quantitative Lipschitz inverse mapping theorem (cf. [134, 285]).

Lemma 4.1. *Assume that Y and Z are Banach spaces, $H \in C^1(Y, Z)$, and $H'(y_0)$ is a homeomorphism for some $y_0 \in Y$. Let k, σ, and $\delta > 0$ be three constants such that the following estimates hold:*

$$\|H'(y) - H'(y_0)\| \leq k < \sigma \leq \frac{1}{\|H'(y_0)^{-1}\|} \quad \text{for any } y \in \mathcal{U}_Y(y_0; \delta),$$

$$\|H(y_0)\| \leq (\sigma - k)\delta.$$

Then H has a unique zero $\bar{y} \in \mathcal{U}_Y(y_0; \delta)$ and the following inequalities are satisfied:

$$\|H'(y)^{-1}\| \leq \frac{1}{\sigma - k} \qquad \forall y \in \mathcal{U}_Y(y_0; \delta),$$

$$\|y_1 - y_2\| \leq \frac{1}{\sigma - k} \|H(y_1) - H(y_2)\| \quad \forall y_1, y_2 \in \mathcal{U}_Y(y_0; \delta).$$

$(4.1.19)$

4.2 Discretization of Operators and Semigroups

The following assertion is a simple consequence of Definition 1.2 (see [271, 285]).

Lemma 4.2. *Let B_n and B be bounded linear operators. Then the convergence*

$$B_n \xrightarrow{\mathcal{PP}} B$$

is equivalent to the boundedness of $\|B_n\|$ and the condition

$$B_n p_n x \xrightarrow{\mathcal{P}} Bx$$

for all $x \in E$. If this convergence holds, then for any compact set $K \subset E$ we have

$$\sup_{x \in K} \left\| B_n p_n x - p_n B x \right\| \to 0 \quad as \ n \to \infty. \tag{4.2.1}$$

Proof. For the first statement, we refer the reader to [271, 285]. The proof of (4.2.1) is by contradiction. Assume that $\|B_n p_n x^n - p_n B x^n\| \geq \varepsilon > 0$ for some sequence $x^n \in K$, $n \in \mathbb{N}$, and some $\varepsilon > 0$. Then, taking a subsequence x^n, $n \in \mathbb{N}' \subset \mathbb{N}$, such that $x^n \to x$ for some $x \in K$, we arrive at the contradiction:

$$\left\| B_n p_n x^n - p_n B x^n \right\| \leq \|B_n\| \left\| p_n(x^n - x) \right\|$$
$$+ \left\| B_n p_n x - p_n B x \right\| + \left\| p_n B(x - x^n) \right\| \to 0$$

for $n \in \mathbb{N}'$. $\qquad \square$

As a simple consequence, we obtain the uniform convergence on compact sets (the proof is similar to the proof of (4.2.1)).

Corollary 4.1. *Under the assumptions (A) and (B$_1$) of Theorem 1.3, for any compact set $K \subset E$ we have*

$$\max_{u^0 \in K} \max_{\eta \in \Sigma(\theta, \mu)} \left\| \left(\exp(\eta A_n) p_n - p_n \exp(\eta A) \right) u^0 \right\| \to 0 \quad as \ n \to \infty. \tag{4.2.2}$$

The semidiscrete approximation of (4.1.1) is the set of the following Cauchy problems in the Banach spaces E_n:

$$
\begin{aligned}
u_n'(t) &= A_n u_n(t) + g_n(t), \quad t \in [0, T], \\
u_n(0) &= u_n^0,
\end{aligned}
\tag{4.2.3}
$$

where the operators A_n generate C_0-semigroups, (A_n, A) are compatible, $u_n^0 \xrightarrow{\mathcal{P}} u^0$, and $g_n(\cdot) \xrightarrow{\mathcal{P}} g(\cdot)$ in an appropriate sense. For a typical semidiscretization, it is natural to assume that conditions similar to (A) and (B$_1$) are satisfied.

Comparing the condition $\Delta_{cc} \neq \varnothing$ with the regular compatibility (ii) (see Definition 3.3), we see that the first of them implies the second. Moreover, by [120], i.e., by Theorems 3.7 and 3.8, we have

$$
\Delta_{cc} \neq \varnothing \quad \Longrightarrow \quad \Delta_{cc} = \Delta_c \cap \rho(A) \text{ and } \Delta_r = \mathbb{C}.
$$

4.2.1 *Estimates for the linear case*

Let Λ be a compact subset of $\rho(A)$. By Lemma 3.5, there is a constant $n_\Lambda > 0$ such that $\Lambda \subset \rho(A_n)$ for all $n \geq n_\Lambda$ and

$$
\sup_{\substack{\lambda \in \Lambda, \\ n \geq n_\Lambda}} \left\| (\lambda I_n - A_n)^{-1} \right\| < \infty.
$$

Consider the contour G consisting of the set $\{\lambda : \operatorname{Re} \lambda = -\tilde{\omega}\}$ and the part of the boundary $\partial \Sigma(\pi - \phi, \omega_3)$ (see Fig. 3.3, p. 61). We have the representation

$$
\exp(t A_n) = \frac{1}{2\pi i} \int_G e^{\lambda t} (\lambda I_n - A_n)^{-1} d\lambda.
$$

Using the estimate (3.5.12) of the resolvents $\left\| (\lambda I_n - A_n)^{-1} \right\|$ and the uniform estimate on G, we obtain the estimate (3.5.11) for $t > 0$ (cf. (4.1.12)).

Next, we introduce the operators

$$
p_n^\theta = (-A_n)^{-\theta} p_n (-A)^\theta \in B(E^\theta, E_n^\theta)
\tag{4.2.4}
$$

and show that they possess the property (1.3.1) for the spaces E^θ and E_n^θ. Then we write $x_n \xrightarrow{\mathcal{P}^\theta} x$ if and only if $\left\| x_n - p_n^\theta x \right\|_{E_n^\theta} \to 0$ as $n \to \infty$. Obviously, we have

$$
\left\| x_n - p_n^\theta x \right\|_{E_n^\theta} = \left\| (-A_n)^\theta x_n - p_n (-A)^\theta x \right\|_{E_n},
$$

$$
\left\| p_n^\theta x \right\|_{E_n^\theta} = \left\| p_n (-A)^\theta x \right\|_{E_n} \to \left\| (-A)^\theta x \right\|_{E_n} = \left\| x \right\|_{E^\theta} \quad \text{as } n \to \infty
$$

for any $x \in D((-A)^\theta)$, so that (1.3.1) is satisfied.

For the nonlinear case, we need a theorem on the uniform convergence for linear inhomogeneous problems with compact data.

Theorem 4.1. *Let $A : D(A) \subset E \to E$ be a closed operator satisfying the condition (1.1.3) (in particular, A generates an exponentially decreasing semigroup). For the approximate system (4.2.3), assume that $\Delta_{cc} \neq \varnothing$ and let Conditions (A) and (B$_1$) of Theorem 1.3 be fulfilled. Let $K_1 \subset E^{\theta}$ and $K_2 \subset E$ be compact sets. Fix $T \in (0; \infty]$. Then for any $\varepsilon > 0$, there exist $n_1 = n_1(\varepsilon) \in \mathbb{N}$ and $\delta = \delta(\varepsilon) > 0$ such that for all functions $g(\cdot) \in C([0, T]; K_2)$ and $g_n(\cdot) \in C([0, T]; E_n)$, $n \in \mathbb{N}$, satisfying the condition*

$$\left\| g_n(t) - p_n g(t) \right\|_{E_n} \leq \delta \quad \text{for } t \in [0, T] \text{ and } n \geq n_1,$$

the following estimate holds for any solution $u(t)$ of the problem (4.1.1) with $u(0) = u^0 \in K_1$ and any solution $u_n(t)$ of the problem (4.2.3) with $u_n(0) = p_n^{\theta} u^0$:

$$\left\| u_n(t) - p_n^{\theta} u(t) \right\|_{E_n^{\theta}} \leq \epsilon \quad \text{for all } n \geq n_1 \text{ and } 0 \leq t \leq T. \qquad (4.2.5)$$

Proof. Let $u(t)$, $u(0) = u^0$, and $u_n(t; u_n^{\theta})$, $u_n^{\theta} = p_n^{\theta} u^0$, denote mild solutions of the problems (4.1.1) and (4.2.3), respectively, and let g and g_n satisfy the conditions of the theorem. Then the following relation holds:

$$u(t; u^0) = \exp(tA)u^0 + \int_0^t \exp\big((t - s)A\big)g(s)ds, \qquad t \in [0, T],$$

$$u_n(t; u_n^{\theta}) = \exp(tA_n)u_n^{\theta} + \int_0^t \exp\big((t - s)A_n\big)g_n(s)ds, \quad t \in [0, T].$$

By Theorem 3.16, we have

$$\left\| \exp(tA_n) \right\| \leq Me^{-\omega t}, \quad \omega > 0, \quad t \geq 0.$$

Therefore, for any $\epsilon > 0$, there is T_{ϵ} such that

$$\left\| \big(\exp(tA_n)p_n^{\theta} - p_n^{\theta} \exp(tA) \big)u^0 \right\|_{E_n^{\theta}} \leq Me^{-\omega t}\|u^0\|_{E^{\theta}} \leq \epsilon, \quad t \geq T_{\epsilon}, \ (4.2.6)$$

and by Corollary 4.1, there is a number $n(T_{\epsilon})$ such that

$$\left\| (-A_n)^{\theta} \big(\exp(tA_n)p_n^{\alpha} - p_n^{\alpha} \exp(tA) \big)u^0 \right\|_{E_n} \leq \epsilon, \quad t \in [0, T_{\epsilon}], \ n \geq n(T_{\epsilon}).$$
$$(4.2.7)$$

We note that in the case where $T_{\epsilon} > T$, we merely need a compactness argument. A similar remark applies to the following estimates of integrals. For simplicity, we extend $g(t)$ and $g_n(t)$ as constants for $t \geq T$ and, on the interval $[0, T]$, choose these functions with the indicated properties.

Consider the E_n^θ-norm of the difference of integrals, i.e.,

$$(-A_n)^\theta \int_0^t \Big(\exp\big((t-s)A_n\big)g_n(s) - p_n^\theta \exp\big((t-s)A\big)g(s) \Big)ds$$

$$= (-A_n)^\theta \int_0^t \exp\big((t-s)A_n\big)\Big(g_n(s) - p_ng(s)\Big)ds$$

$$+ (-A_n)^\theta \int_0^t \Big(\exp\big((t-s)A_n\big)p_n - p_n^\theta \exp\big((t-s)A\big) \Big)g(s)\,ds.$$

$$(4.2.8)$$

The first term can be estimated as follows:

$$(-A_n)^\theta \Big(\int_0^{t_1} + \int_{t_1}^t \Big)$$

$$= (-A_n)^\theta \int_0^{t_1} \exp(sA_n)\Big(g_n(t-s) - p_ng(t-s)\Big)ds$$

$$+ (-A_n)^\theta \int_0^{t-t_1} \exp\big((t_1+\eta)A_n\big)\Big(g_n(t-t_1-\eta) - p_ng(t-t_1-\eta)\Big)d\eta.$$

By (3.5.11), the second part can be made less than ε uniformly in n by taking sufficiently large t_1. Then the first part with finite t_1 can be made small by the majorant term $\big\|g_n(t) - p_ng(t)\big\|_{E_n} \le \delta$ as $n \ge n_1$. The second term in (4.2.8) for any $0 < t_1 \le t$ can be rewritten as follows:

$$(-A_n)^\theta \int_0^t \Big(\exp(\eta A_n)p_n^\theta - p_n^\theta \exp(\eta A) \Big)g(t-\eta)d\eta$$

$$= (-A_n)^\alpha \int_0^{t_1} \Big(\exp(\eta A_n)p_n - p_n^\theta \exp(\eta A)B \Big)g(t-\eta)d\eta$$

$$+ (-A_n)^\theta \int_0^{t-t_1} \Big(\exp\big((t_1+\eta)A_n\big)p_n$$

$$- p_n^\theta \exp\big((t_1+\eta)A\big) \Big)g(t-t_1-\eta)d\eta.$$

Again, we first choose t_1 to make the second term small uniformly in n. Then by Corollary 4.1 the first term converges to 0 for finite t_1 since $g(\xi)$ belongs to the compact set K_2. The second term can be decomposed into two parts:

$$(-A_n)^\theta \exp(t_1 A_n) \int_0^{t-t_1} \Big(\exp(\eta A_n)p_n - p_n^\theta \exp(\eta A) \Big)g(t-t_1+\eta)d\eta,$$

$$(-A_n)^\theta \Big(\exp(t_1 A_n)p_n^\theta - p_n^\theta \exp(t_1 A) \Big) \int_0^{t-t_1} \exp(\eta A)g(t-t_1+\eta)d\eta.$$

Both parts can be made small by an appropriate choice of t_1, using (3.5.11), (4.1.12), and uniform bounds of the integrals. $\qquad\square$

4.2.2 Estimates for the nonlinear case

In the Banach spaces E_n^θ, consider the following family of Cauchy problems:

$$u_n'(t) = A_n u_n(t) + f_n(u_n(t)), \quad t \geq 0,$$
$$u_n(0) = u_n^0 \in E_n^\theta, \tag{4.2.9}$$

where $u_n^0 \xrightarrow{\mathcal{P}^\theta} u^0$ and the operators (A_n, A) are compatible. We assume that the nonlinear maps $f_n(\cdot) : E_n^\theta \to E_n$ possess the following properties:

(F2) the mappings $f_n(\cdot)$ are continuously differentiable in $\mathcal{U}_{E_n^\theta}(p_n^\theta u^*, \rho)$; moreover, if $x_n \in \mathcal{U}_{E_n^\theta}(p_n^\theta u^*, \rho)$ and $x_n \xrightarrow{\mathcal{P}^\theta} x$, then $f_n(x_n) \xrightarrow{\mathcal{P}} f(x)$ and $f_n'(x_n) \xrightarrow{\mathcal{P}^\theta \mathcal{P}} f'(x)$;

(F3) for any $\epsilon > 0$, there exists $\delta > 0$ such that

$$\left\| f_n'(w_n) - f_n'(z_n) \right\|_{B(E_n^\theta, E_n)} \leq \epsilon \quad \text{if } \|w_n - z_n\|_{E_n^\theta} \leq \delta$$

for all $w_n, z_n \in \mathcal{U}_{E_n^\theta}(p_n^\theta u^*; \rho)$.

Under the above assumptions, a mild solution of (4.2.9) exists on a certain maximal interval $[0, \tau)$ in $\mathcal{U}_{E_n^\theta}(p_n^\theta u^*, \rho)$ (see [129, 295]); we denote it by $u_n(\cdot) = S_n(\cdot)u_n^0 : \mathbb{R}^+ \to E_n$. The nonlinear semigroup $S_n(\cdot)$ satisfies the formula of variation of constants:

$$S_n(t)u_n^0 = \exp(tA_n)u_n^0 + \int_0^t \exp\big((t-s)A_n\big)f_n(S_n(s)u_n^0)ds, \quad t \in [0, \tau). \tag{4.2.10}$$

We consider the family of nonlinear problems (3.3.1) and introduce the notation $\mathcal{E}_n = \big\{ u_n^* \in D(A_n) : \ A_n u_n^* + f_n(u_n^*) = 0 \big\}$.

From now on, we consider a hyperbolic point u^* and the corresponding fixed points $u_n^* \xrightarrow{\mathcal{P}^\theta} u^*$ from Proposition 3.5. Near the equilibrium u_n^*, we set $u_n(t) = u_n^* + v_n(t)$; then the problem (4.2.9) takes the form

$$v_n'(t) = A_{u_n^*,n} v_n(t) + F_{u_n^*,n}(v_n(t)), \quad v_n(0) = v_n^0, \quad t \geq 0, \tag{4.2.11}$$

where

$$A_{u_n^*,n} = A_n + f_n'(u_n^*),$$
$$F_{u_n^*,n}(w_n) = f_n(v_n(t) + u_n^*) - f_n(u_n^*) - f_n'(u_n^*)w_n.$$

We decompose E_n^θ using the spectral projectors

$$P_n(\sigma_n^+) := P_n(\sigma_n^+, A_{u_n^*,n}) := \frac{1}{2\pi i} \int_{\partial U(\sigma_n^+)} \big(\zeta I_n - A_{u_n^*,n}\big)^{-1} d\zeta, \tag{4.2.12}$$

where $\partial U(\sigma_n^+)$ is the boundary of the domain

$$\{\lambda \in \mathbb{C} : \ \mathrm{Re}\,\lambda \geq 0, \ \lambda \notin \Sigma(\pi - \phi, \omega_2)\}$$

with some $0 < \theta < \pi/2$ and ω_2 given by (B$_1$) for the operator $A_{u_n^*,n}$ (cf. Theorem 3.16). Note that the part of $\partial U(\sigma_n^+)$ on $i\mathbb{R}$ does not intersect $\sigma(A_{u_n^*,n})$ due to Lemma 3.5. In particular, this implies that the fixed points u_n^* are hyperbolic.

We denote by σ_n^+ the part of $\sigma(A_{u_n^*,n})$ that lies inside the contour $\partial U(\sigma_n^+)$. From the representations (3.1.5) and (4.2.12) we obtain

$$P_n(\sigma_n^+) \xrightarrow{\ \mathcal{PP}\ } P(\sigma^+) \quad \text{compactly as } n \to \infty. \tag{4.2.13}$$

To see this, we first modify the contour in (3.1.5) so that it coincides with the contour in (4.2.12) and then use the convergence

$$(\zeta I_n - A_{u_n^*,n})^{-1} \xrightarrow{\ \mathcal{PP}\ } (\zeta I - A_{u^*})^{-1} \quad \text{for } \zeta \in \partial U(\sigma^+)$$

(see Theorem 3.8) and the fact that the convergence

$$(\zeta I_n - A_{u_n^*,n})^{-1} p_n x \xrightarrow{\ \mathcal{P}\ } (\zeta I - A_{u^*})^{-1} x$$

is uniform for $\zeta \in \partial U(\sigma^+)$. Moreover, the condition $\Delta_{cc} \neq \varnothing$ implies the compact convergence of projectors and, therefore, by Theorems 3.8 and 3.6, $\dim P_n(\sigma_n^+) = \dim P(\sigma^+)$ for $n \geq n_0$.

Applying Theorem 3.16 to the semigroup $\exp(tA_{u_n^*,n})$, we obtain constants M_2 and $\tilde{\beta} > 0$ such that

$$\left\| \exp\left(tA_{u_n^*,n}\right)v_n \right\|_{E_n^\theta} \leq \begin{cases} M_2 \dfrac{e^{-\tilde{\beta}t}}{|t|^\theta}\|v_n\|_{E_n}, & t \geq 0, \ v_n \in (I_n - P_n(\sigma_n^+))E_n, \\[3mm] M_2 \dfrac{e^{\tilde{\beta}t}}{|t|^\theta}\|v_n\|_{E_n}, & t \leq 0, \ v_n \in P_n(\sigma_n^+)E_n. \end{cases}$$

$$\tag{4.2.14}$$

Similarly to (4.1.9), we consider for any $v_n^-, v_n^+ \in \mathcal{U}_{E_n^\theta}(0; \rho)$ the boundary-value problem

$$\begin{aligned} v_n'(t) &= A_{u_n^*,n}v_n(t) + F_{u_n^*,n}(v_n(t)), \quad 0 \leq t \leq T, \\ (I_n - P_n(\sigma_n^+))v_n(0) &= (I_n - P_n(\sigma_n^+))v_n^-, \\ P_n(\sigma_n^+)v_n(T) &= P_n(\sigma_n^+)v_n^+, \end{aligned} \tag{4.2.15}$$

where the case $T = \infty$ is included in the usual way. Using Condition (F2), we see that a mild solution of the problem (4.2.15) satisfies for $0 \leq t \leq T$ the equation

$$\begin{aligned} v_n(t) &= \exp\left((t-T)A_{u_n^*,n}\right)P_n(\sigma_n^+)v_n^+ + \exp\left(tA_{u_n^*,n}\right)\left(I_n - P_n(\sigma_n^+)\right)v_n^- \\ &\quad + \int_0^T \Gamma_{T,n}(t,s)F_{u_n^*,n}(v_n(s))ds, \end{aligned} \tag{4.2.16}$$

where $\Gamma_{T,n}(t,s)$ is the Green function

$$\Gamma_{T,n}(t,s) = \begin{cases} \exp\left((t-s)A_{u^*,n}\right)\left(I_n - P_n(\sigma_n^+)\right), & 0 \le s \le t \le T, \\ \exp\left((t-s)A_{u^*,n}\right)P_n(\sigma_n^+), & 0 \le t < s \le T. \end{cases}$$

(4.2.17)

In the case where $T = \infty$, the term v_n^+ in (4.2.16) vanishes. The following assertion is an analog of Proposition 4.1.

Proposition 4.2 (see [56]). *Let the above assumptions on the operators A and A_n and Conditions (F2) and (F3) be fulfilled. Then there exists $\tilde{\rho} > 0$ such that for any $0 < \tilde{\rho}_2 \le \tilde{\rho}$, one can find $0 < \tilde{\rho}_1 \le \tilde{\rho}_2$ with the following property: Eq. (4.2.16) has a unique solution*

$$v_n(\cdot) = v_n(v_n^-, v_n^+, \cdot) \in C([0,T]; \mathcal{U}_{E_n^\theta}(0; \tilde{\rho}_2))$$

for all $v_n^-, v_n^+ \in \mathcal{U}_{E_n^\theta}(0; \tilde{\rho}_1)$ and all $0 < T \le \infty$. If $T = \infty$, then

$$\|v_n(t)\|_{E_n^\theta} \to 0 \quad \text{as } t \to \infty.$$

Proof. We repeat the proof of Proposition 4.1 for the space of continuous bounded functions $C([0,T]; E_n^\theta)$ with the operators $A_{u_n^*,n}$, $F_{u_n^*,n}(\cdot)$, $P_n(\sigma_n^+)$, and $G_n(v_n^-, v_n^+; \cdot)$ defined by the right-hand side of (4.2.16). Note that Condition (F3) guarantees that the estimates in (4.2.14) can hold with constants independent of n and, therefore, Lemma 4.1 is valid with uniform data. Moreover, the estimates (4.1.17) and (4.1.18) hold uniformly in n. From (4.1.19) we find a constant $C^* > 0$ such that for any two $v_n, w_n \in C([0,T]; \mathcal{U}_{E_n^\theta}(0; \tilde{\rho}_2))$ we have the estimate

$$\|v_n - w_n\| \le C^* \left\| v_n - G_n\left(v_n^-, v_n^+; v_n\right) - \left(w_n - G_n(v_n^-, v_n^+; w_n)\right) \right\|, \quad (4.2.18)$$

where $\|\cdot\| = \|\cdot\|_{C([0,T]; E_n^\theta)}$. $\qquad\square$

4.3 Shadowing in Space

In addition to p_n and p_n^θ, we need discretizing maps that are adapted to the hyperbolic splitting. First, we note that the spectral projectors $P = P(\sigma^+)$ and $P_n = P_n(\sigma_n^+)$ are finite-dimensional and $P_n \xrightarrow{\mathcal{PP}} P$ compactly (see (4.2.13)). Then we define the discretizing maps $\tilde{p}_n : E \mapsto E_n$ by the rule

$$\tilde{p}_n : E \to E_n, \quad \tilde{p}_n = P_n p_n P + (I_n - P_n)p_n(I - P), \quad (4.3.1)$$

and

$$\tilde{p}_n^\theta : E^\theta \to E_n^\theta, \tag{4.3.2}$$

$$\tilde{p}_n^\theta x = \begin{cases} (A_{u_n^*,n})^{-\theta} P_n p_n (A_{u^*})^\theta P x, & x \in PE^\theta, \\ (-A_{u_n^*,n})^{-\theta} (I_n - P_n) p_n (-A_{u^*})^\theta (I - P) x, & x \in (I - P)E^\theta. \end{cases}$$

Note that the spectra of the operators

$$A_{u^*} = A + f'(u^*), \quad A_{u_n^*,n} = A_n + f_n'(u_n^*)$$

are partitioned in such a way that the fractional powers of the operators are well defined.

Proposition 4.3 (see [56]). *The system $\{\tilde{p}_n\}$ is equivalent to the system $\{p_n\}$ on E and the system $\{\tilde{p}_n^\theta\}$ is equivalent to the system $\{p_n^\theta\}$ on E^θ. In particular, we have*

$$\left\| \tilde{p}_n \right\|_{B(E,E_n)} \leq \tilde{C}, \quad \left\| \tilde{p}_n^\theta \right\|_{B(E^\theta,E_n^\theta)} \leq \tilde{C}_\theta \quad \text{for all } n \in \mathbb{N}, \tag{4.3.3}$$

$$\sup_{x \in K_2} \left\| (p_n - \tilde{p}_n)x \right\| \to 0, \quad \sup_{x \in K_1} \left\| (p_n^\theta - \tilde{p}_n^\theta)x \right\| \to 0 \quad \text{as } n \to \infty \tag{4.3.4}$$

for compact sets $K_2 \subset E, K_1 \subset E^\theta$.

Proof. From the equality

$$(\tilde{p}_n - p_n)x = 2(P_n p_n - p_n P)P x - (P_n p_n - p_n P)x$$

we find that the system $\{\tilde{p}_n\}$ is equivalent to $\{p_n\}$ on E. In the θ-case, the following chain of equalities holds on $(I - P)E^\theta$:

$$\begin{aligned}
p_n^\theta - \tilde{p}_n^\theta &= (-A_n)^{-\theta}(p_n - \tilde{p}_n)(-A)^\theta + (-A_n)^{-\theta}\tilde{p}_n(-A)^\alpha \\
&\quad - (-A_{u_n^*,n})^{-\theta}\tilde{p}_n(-A_{u^*})^\theta \\
&= (-A_n)^{-\theta}(p_n - \tilde{p}_n)(-A)^\theta \\
&\quad + \left((-A_n)^{-\theta} - (-A_{u_n^*,n})^{-\theta} \right)\tilde{p}_n(-A)^\theta \\
&\quad + (-A_{u_n^*,n})^{-\theta}\tilde{p}_n \left((-A)^\theta - (-A_{u^*})^\theta \right) \\
&= (-A_n)^{-\theta}(p_n - \tilde{p}_n)(-A)^\theta \\
&\quad + \left((-A_n)^{-\theta} - (-A_{u_n^*,n})^{-\theta} \right)\tilde{p}_n(-A)^\theta \\
&\quad + (-A_{u_n^*,n})^{-\theta}\tilde{p}_n \left((-A)^\theta - (-A_{u^*})^\theta \right) \\
&= (-A_n)^{-\theta}(p_n - \tilde{p}_n)(-A)^\theta \\
&\quad + (-A_{u_n^*,n})^{-\theta}\left((-A_{u_n^*,n})^\theta (-A_n)^{-\theta} - I_n \right)\tilde{p}_n(-A)^\theta \\
&\quad - (-A_{u_n^*,n})^{-\theta}\tilde{p}_n \left((-A_{u^*})^\theta (-A)^{-\theta} - I \right)(-A)^\theta
\end{aligned}$$

$$
= (-A_n)^{-\theta}(p_n - \tilde{p}_n)(-A)^{\theta}(-A_{u_n^*,n})^{-\theta}
$$
$$
+ \Big(\big(((-A_{u_n^*,n})^{\theta}(-A_n)^{-\theta} - I_n)\tilde{p}_n
$$
$$
- \tilde{p}_n ((-A_{u^*})^{\theta}(-A)^{-\theta} - I) \big) \Big) (-A)^{\theta}.
$$

At the last step, we show that

$$
(-A_{u_n^*,n})^{\theta}(-A_n)^{-\theta} \xrightarrow{\;\mathcal{PP}\;} (-A_{u^*})^{\theta}(-A)^{-\theta}.
$$

For this purpose, following [129], we consider the formula

$$
(-A_{u^*})^{-\theta} - (-A)^{-\theta}
$$
$$
= \frac{\sin(\pi\theta)}{\pi} \int_0^{\infty} z^{-\theta} \Big((zI + A_{u^*})^{-1} - (zI + A)^{-1} \Big) dz
$$
$$
= \frac{\sin(\pi\theta)}{\pi} \int_0^{\infty} z^{-\theta} \Big((zI + A_{u^*})^{-1}
$$
$$
\times \big(f'(u^*)(-A)^{-\theta} \big)(-A)^{\theta}(zI + A)^{-1} \Big) dz. \qquad (4.3.5)
$$

Since

$$
\big\| (-A)^{\theta}(zI + A)^{-1} \big\| = O\big(|z|^{\theta-1}\big) \quad \text{as } z \to \infty,
$$

the integral will converge even if we apply $(-A_{u^*})^{\theta}$. Therefore, the convergence

$$
(-A_{u_n^*,n})^{\theta}(-A_n)^{-\theta} \xrightarrow{\;\mathcal{PP}\;} (-A_{u^*})^{\theta}(-A)^{-\theta}
$$

follows from

$$
f_n'(u_n^*)(-A_n)^{-\theta} \xrightarrow{\;\mathcal{PP}\;} f'(u^*)(-A)^{-\theta}
$$

and the Lebesgue dominated convergence theorem. The uniform convergence on compact sets follows as in Lemma 4.2. $\qquad \square$

Remark 4.2. It is easy to see that $(-A)^{\theta}(-A_{u^*})^{-\theta}$ is a bounded operator and

$$
(-A_n)^{\theta}(-A_{u_n^*,n})^{-\theta} \xrightarrow{\;\mathcal{PP}\;} (-A)^{\theta}(-A_{u^*})^{-\theta}.
$$

Indeed,

$$
\big\| (-A)^{\theta}(zI - A_{u^*})^{-1} \big\| = O\big(|z|^{\theta-1}\big) \quad \text{as } z \to \infty;
$$

therefore, interchanging A and A_{u^*} in (4.3.5), we get the boundedness and convergence in the same way as in Proposition 4.3.

Proposition 4.4. *Let* $0 < \theta < \gamma \leq 1$ *and Conditions* (F1)–(F3) *be fulfilled. Then*

$$\sup_{w \in \mathcal{U}_{E^\gamma}(0;\rho)} \left\| F_{u_n^*,n}(\tilde{p}_n^\theta w) - \tilde{p}_n F_{u^*}(w) \right\|_{E_n} \to 0 \quad as \ n \to \infty. \quad (4.3.6)$$

Proof. First, we note that

$$F_{u_n^*,n}(\tilde{p}_n^\theta w) - \tilde{p}_n F_{u^*}(w) = f_n(u_n^* + \tilde{p}_n^\theta w) - \tilde{p}_n f(u^* + w) -$$
$$- \left(f_n(u_n^*) - \tilde{p}_n f(u^*) + f_n'(u_n^*)\tilde{p}_n^\theta w - \tilde{p}_n f'(u^*)w \right)$$

and observe that by Proposition 4.3 we can replace the maps $\tilde{p}_n$ and $\tilde{p}_n^\theta$ by p_n and p_n^θ, respectively. From Proposition 3.5 and Condition (F2) we obtain

$$f(u_n^* + p_n^\theta w) \xrightarrow{\mathcal{P}} f(u^* + w),$$
$$f_n(u_n^*) \xrightarrow{\mathcal{P}} f(u^*),$$
$$f_n'(u_n^*) \xrightarrow{\mathcal{P}^\theta \mathcal{P}} f'(u^*).$$

Following the proof of the relation (4.2.1), we then show that the convergence is uniform on compact sets in E^θ. □

4.3.1 *Shadowing in space and the condition* $\Delta_{cc} \neq \varnothing$

The first result concerns approximations of orbits of the general evolution equation (4.1.5) by appropriate orbits of the "spatially discretized" system (4.2.9).

Theorem 4.2 (see [56]). *Let A be the generator of an exponentially decreasing, analytic C_0-semigroup and $0 \leq \theta < \gamma < 1$. For the discretized system (4.2.9) assume that the linear parts satisfy the conditions $\Delta_{cc} \neq \varnothing$ and* (B$_1$) *and the nonlinear parts satisfy Conditions* (F1), (F2), *and* (F3). *Then there exists $\rho_0 > 0$ possessing the following property: for any $\varepsilon_0 > 0$, there is a number $n_0 = n_0(\varepsilon_0) \in \mathbb{N}$ such that for any mild solution $u(t)$ of (4.1.5) satisfying the condition $u(t) \in \mathcal{U}_{E^\gamma}(u^*, \rho_0)$, $0 \leq t \leq T$, for some $0 < T \leq \infty$, there exist initial values $u_n^0 \in E_n^\theta$, $n \geq n_0$, such that the mild solution $u_n(t; u_n^0)$ of (4.2.9) exists on $[0, T]$ and satisfies the inequality*

$$\sup_{0 \leq t \leq T} \left\| p_n^\theta u(t) - u_n(t; u_n^0) \right\|_{E_n^\theta} \leq \varepsilon_0 \quad \forall n \geq n_0(\varepsilon). \quad (4.3.7)$$

Proof. We will collect the conditions on n_0 and ρ_0 during the proof. Let $u(t)$, $0 \leq t \leq T$, be a mild solution of (4.1.5) such that $u(t) \in \mathcal{U}_{E^\gamma}(u^*, \rho_0)$

for all $0 \leq t \leq T$. Then $v(t) := u(t) - u^* \in \mathcal{U}_{E^\gamma}(0, \rho_0)$ is a solution of (4.1.10) such that

$$v^- = (I - P)v(0), \quad v^+ = Pv(T).$$

By the choice of norms (4.1.8), we have

$$\|v^-\|_{E^\gamma} \leq \rho_0, \quad \|v^+\|_{E^\gamma} \leq \rho_0.$$

We apply Proposition 4.1 replacing θ by γ and setting $\hat{\rho}_2 = \hat{\rho}$. We also require $\rho_0 \leq \hat{\rho}_1$, which implies $\rho_0 \leq \hat{\rho}_2$. Due to the uniqueness of a solution in $C([0,T]; \mathcal{U}_{E^\gamma}(0, \hat{\rho}_2))$, the solution $v(v^-, v^+, \cdot)$ from Proposition 4.1 satisfies the condition $v(t) = v(v^-, v^+, t)$, $0 \leq t \leq T$. Next, we define the discrete boundary values

$$v_n^- = \tilde{p}_n^\theta v^-, \quad v_n^+ = \tilde{p}_n^\theta v^+. \tag{4.3.8}$$

By the definition (4.3.2) we have

$$v_n^- = \tilde{p}_n^\theta v^- = (I_n - P_n)\tilde{p}_n^\theta v^-, \quad v_n^+ = \tilde{p}_n^\theta v^+ = P_n \tilde{p}_n^\theta v^+,$$

and from (4.3.3) we have

$$\|v_n^\pm\|_{E_n^\theta} \leq \tilde{C}_\theta \|v^\pm\|_{E^\theta} \leq \tilde{C}_\theta \rho_0. \tag{4.3.9}$$

Then we apply Proposition 4.2 with $\tilde{\rho}_2 = \tilde{\rho}$ and require that $\tilde{C}_\theta \rho_0 \leq \tilde{\rho}_1(\tilde{\rho}_2)$. Taking the corresponding unique solutions of (4.2.16)

$$v_n(\cdot) = v_n(v_n^-, v_n^+, \cdot) \in C([0,T]; \mathcal{U}_{E_n^\theta}(0, \tilde{\rho}_2)) \tag{4.3.10}$$

we claim that

$$u_n^0 = u_n^* + v_n(0), \quad u_n(t) = v_n(t) + u_n^* \tag{4.3.11}$$

satisfy the assertion of the theorem. We require $\tilde{C}_\theta \rho_0 \leq \tilde{\rho}_1$, so that we can apply (4.2.18) to $v_n(t)$ and $w_n(t) = \tilde{p}_n^\theta v(t)$. We obtain

$$\sup_{0 \leq t \leq T} \left\| v_n(t) - \tilde{p}_n^\theta v(t) \right\|_{E_n^\theta} \leq C^* \sup_{0 \leq t \leq T} \left\| \eta_n^-(t) + \eta_n^+(t) \right\|_{E_n^\theta}; \tag{4.3.12}$$

where the terms on the right-hand side are given by the formulas

$$\eta_n^-(t) = \tilde{p}_n^\theta \exp(tA_{u^*})(I - P)v^- - \exp(tA_{u_n^*,n})(I_n - P_n)v_n^-$$

$$+ \tilde{p}_n^\theta \int_0^t \exp((t-s)A_{u^*})(I - P)F_{u^*}(v(s))ds$$

$$- \int_0^t \exp((t-s)A_{u_n^*,n})(I_n - P_n)F_{u_n^*,n}(\tilde{p}_n^\theta v(s))ds$$

and

$$\eta_n^+(t) = \tilde{p}_n^\theta \exp\left((t-T)A_{u^*}\right)Pv^+ - \exp\left((t-T)A_{u_n^*,n}\right)P_n v_n^+$$
$$+ \tilde{p}_n^\theta \int_t^T \exp\left((t-s)A_{u^*}\right)PF_{u^*}(v(s))ds$$
$$- \int_t^T \exp\left((t-s)A_{u_n^*,n}\right)P_n F_{u_n^*,n}\left(\tilde{p}_n^\theta v(s)\right)ds.$$

We estimate η_n^+ using Theorem 4.1 with the following settings:

$$\tilde{E} = (I-P)E, \quad D(\tilde{A}) = D(A) \cap \tilde{E}, \quad \tilde{A} = A_{u^*},$$
$$g(t) = (I-P)F_{u^*}(v(t)), \quad g_n(t) = (I_n - P_n)F_{u_n^*,n}\left(\tilde{p}_n^\theta v(t)\right),$$
$$K_1 = \mathcal{U}_{E^\gamma}(0, \rho_0), \quad K_2 = \left\{(I-P)F_{u^*}(w) : \ w \in \mathcal{U}_{E^\gamma}(0, \rho_0)\right\},$$
$$\tilde{\varepsilon} = \frac{\varepsilon_0}{4C^*}, \quad u^0 = v^-, \quad \tilde{E}_n = (I_n - P_n)E_n, \quad \tilde{E}_n^\theta = (I_n - P_n)E_n^\theta,$$
$$u_n(0) = \tilde{p}_n^\theta u(0) = \tilde{p}_n^\theta v^-.$$

Note that by the continuity of the function $F_{u^*} : E^\theta \to E$ and the compact embedding $E^\gamma \subset E^\theta$, the set K_1 is compact in E^θ and K_2 is compact in E. We apply the estimate (4.2.5) for $n \geq n_3 = \max(n_1(\tilde{\varepsilon}), n_2)$, where n_2 is chosen by Proposition 4.4 such that for $n \geq n_2$

$$\sup_{0 \leq t \leq T} \left\|g_n(t) - \tilde{p}_n g(t)\right\|_{\tilde{E}_n}$$
$$\leq \sup_{0 \leq t \leq T} \left\|(I-P_n)\left(F_{u_n^*,n}\left(\tilde{p}_n^\theta v(t)\right) - p_n F_{u^*}(v(t))\right)\right\|_{E_n}$$
$$+ \sup_{w \in K_2} \left\|(I-P_n)(p_n P - P_n p_n)F_{u^*}(w)\right\|_{E_n} \leq \delta = \delta(\tilde{\varepsilon}).$$

Therefore, we have

$$\|\eta_n^-\|_{E_n^\theta} \leq \tilde{\varepsilon}$$

and a similar estimate for $\|\eta_n^+\|_{E_n^\theta}$ and $n \geq n_4$. Finally, by (4.3.12) and Proposition 3.5, for some $n_5 \geq n_4$ and all $0 \leq t \leq T$, $n \geq n_5$, we obtain the estimate

$$\left\|u_n(t) - p_n^\theta u(t)\right\|_{E_n^\theta} \leq \left\|v_n(t) - p_n^\theta v(t)\right\|_{E_n^\theta} + \left\|u_n^* - p_n^\theta u^*\right\|_{E_n^\theta}$$
$$\leq \frac{\varepsilon_0}{2} + \frac{\varepsilon_0}{2} = \varepsilon_0.$$

The proof is complete. $\square$

In the following lemma, we approximate vectors of a compact sequence in the discrete spaces by discretizations of continuous elements.

Lemma 4.3. *Let $0 \leq \theta < \gamma < 1$ and $\{v_n^0\}$ be a bounded sequence in E_n^γ. Then for any $\varepsilon > 0$, there is a number $n_0(\varepsilon)$ such that*

$$\inf_{v \in E^\theta} \left\| v_n^0 - \tilde{p}_n^\theta v \right\|_{E_n^\theta} \leq \varepsilon, \quad n \geq n_0(\varepsilon).$$

In addition, if $0 \leq \theta \leq \beta < \gamma < 1$ and the sequence $\{v_n^0\}$ lies in $P_n E_n^\gamma$ and satisfies the condition $\|v_n^0\|_{E_n^\gamma} \leq b$, then there exist $n_1(\varepsilon)$ and a constant $\hat{C} > 0$ such that

$$\inf_{\substack{v \in PE^\beta, \\ \|v\|_{E^\beta} \leq \hat{C}b}} \left\| v_n^0 - p_n^\theta v \right\|_{E_n^\theta} \leq \varepsilon \quad \text{for all } n \geq n_1(\varepsilon). \tag{4.3.13}$$

If, instead of $v_n^0 \in P_n E_n^\gamma$, the condition $v_n^0 \in (I_n - P_n)E_n^\gamma$ is fulfilled, then the inequality (4.3.13) holds with the infimum taken over $v \in (I - P)E^\beta$ and $\|v\|_{E^\beta} \leq \hat{C}b$.

Proof. Proposition 4.3 shows that it suffices to prove the assertion with p_n^θ instead of $\tilde{p}_n^\theta$. Assume that the statement is invalid. Then there exists a sequence $\{v_n^0\}$, a number $\varepsilon > 0$, and a subsequence $\mathbb{N}' \subseteq \mathbb{N}$ such that

$$\left\| (-A_n)^\gamma v_n^0 \right\| \leq b, \quad \left\| v_n^0 - p_n^\theta v \right\|_{E_n^\theta} > \varepsilon$$

for all $n \in \mathbb{N}'$, $v \in E^\theta$. First, note that the sequence $(-A)_n^{-\theta}(-A)_n^\theta v_n^0 = v_n^0$ is $\mathcal{P}$-compact and therefore there exist $\mathbb{N}'' \subseteq \mathbb{N}'$ and $\bar{v} \in E$ such that $v_n^0 \xrightarrow{\mathcal{P}} \bar{v}$ as $n \in \mathbb{N}''$. Since $0 \leq \theta < \gamma < 1$ and

$$(-A_n)^{\theta - \gamma}(-A_n)^\gamma v_n^0 = (-A_n)^\theta v_n^0,$$

we similarly conclude that $(-A_n)^\theta v_n^0$ is $\mathcal{P}$-compact. Thus, there is $\mathbb{N}''' \subseteq \mathbb{N}''$ such that $(-A_n)^\theta v_n^0 \xrightarrow{\mathcal{P}} z \in E$ as $n \in \mathbb{N}'''$ and we have

$$(-A_n)^{-\theta}(-A_n)^\theta v_n^0 \xrightarrow{\mathcal{P}} (-A)^{-\theta} z \quad \text{for } n \in \mathbb{N}'''.$$

On the other hand, $v_n^0 \xrightarrow{\mathcal{P}} \bar{v}$ as $n \in \mathbb{N}''$, which implies $\bar{v} = (-A)^{-\theta} z \in E^\theta$. Finally, we have

$$\left\| v_n^0 - p_n^\theta \bar{v} \right\|_{E_n^\theta} = \left\| (-A_n)^\theta v_n^0 - p_n z \right\|_{E_n} \to 0 \quad \text{for } n \in \mathbb{N}''' \text{ and some } \bar{v} \in E^\theta,$$

a contradiction. To prove the second assertion, we extend the previous argument. First, by Lemma 4.2, for every $0 \leq \lambda < 1$, we have a constant $C_{[\lambda]} > 0$ such that

$$\left\| (-A)^{-\lambda} \right\|_{B(E,E)}, \quad \left\| (-A_n)^{-\lambda} \right\|_{B(E_n,E_n)} \leq C_{[\lambda]} \quad \text{for all } n \in \mathbb{N}. \tag{4.3.14}$$

Let $\hat{C} = C_{[\gamma - \beta]} + 1$. We have seen above that $(-A_n)^\beta v_n^0 \xrightarrow{\mathcal{P}} y \in E$ for $n \in \mathbb{N}'''$ and then $\bar{v} = (-A)^{-\beta} y \in E^\beta$. Using the inclusion $v_n^0 \in P_n E_n^\gamma$ and the convergence of the projectors (4.2.13), we find

$$0 = (I_n - P_n) v_n^0 \xrightarrow{\mathcal{P}} \bar{v} - P\bar{v};$$

hence $\bar{v} \in PE^\beta$. Moreover, for large $n \in \mathbb{N}'''$

$$\|\bar{v}\|_{E^\beta} = \|y\|_E \leq \left\|(-A_n)^{\beta - \gamma}(-A_n)^\gamma v_n^0\right\|_{E_n} + b$$
$$\leq \left(C_{[\gamma - \beta]} + 1\right) b = \hat{C}b,$$

a contradiction with $\|v_n^0 - p_n^\theta v\| > \varepsilon$, where $v \in PE^\beta$, $\|v\|_{E^\beta} \leq \hat{C}b$. $\square$

This lemma will be used for constructing appropriate boundary data for the following inverse shadowing result.

Theorem 4.3 (see [56]). *Let the conditions of Theorem 4.2 hold. Then there exists $\rho_0 > 0$ possessing the following property: for any $\varepsilon_0 > 0$, there is a number $n_0 = n_0(\varepsilon_0) \in \mathbb{N}$ such that for any mild solution $u_n(t)$, $n \geq n_0$, of the problem (4.2.9) satisfying the condition $u_n(t) \in \mathcal{U}_{E_n^\gamma}(u_n^*, \rho_0)$, $0 \leq t \leq T$, for some $0 < T \leq \infty$, there exist initial values $u^{n,0} \in E^\theta$, $n \geq n_0$, such that the mild solution $u(t; u^{n,0})$ of the problem (4.1.5) exists on $[0, T]$ and satisfies the estimate*

$$\sup_{0 \leq t \leq T} \left\| u_n(t) - p_n^\theta u(t; u^{n,0}) \right\|_{E_n^\theta} \leq \varepsilon_0 \quad \forall n \geq n_0(\varepsilon_0). \qquad (4.3.15)$$

Proof. As in Theorem 4.2, we take some $\varepsilon_0 > 0$; the conditions for n_0 and ρ_0 will be specified below. Consider a mild solution $u_n(t)$, $0 \leq t \leq T$, of the problem (4.2.9) that lies in $\mathcal{U}_{E_n^\gamma}(u_n^*, \rho_0)$ and set

$$v_n(t) = u_n(t) - u_n^*, \quad v_n^- = (I_n - P_n) v_n(0), \quad v_n^+ = P_n v_n(T). \qquad (4.3.16)$$

By the uniform boundedness of projectors, for some $C_b \geq 1$ we have

$$\|v_n^-\|_{E_n^\gamma} \leq C_b \rho_0, \quad \|v_n^+\|_{E_n^\gamma} \leq C_b \rho_0. \qquad (4.3.17)$$

We apply Proposition 4.2 to the values $v_n^\pm$ with γ instead of θ and $\tilde{\rho}_2 = \tilde{\rho}$. We also require $C_b \rho_0 \leq \tilde{\rho}_1$ so that $v_n(t) = v_n(v_n^-, v_n^+, t)$, $0 \leq t \leq T$, by the uniqueness of solutions in $C([0, T]; \mathcal{U}_{E_n^\gamma}(0; \tilde{\rho}))$. Now we take $\theta < \beta < \gamma$ and use Lemma 4.3 to construct boundary values $v^{n,-} \in (I - P)E^\beta$ and $v^{n,+} \in PE^\beta$, $n \geq n_1(\varepsilon_0)$, such that

$$\left\| v_n^- - \tilde{p}_n^\theta v^{n,-} \right\|_{E_n^\theta} + \left\| v_n^+ - \tilde{p}_n^\theta v^{n,+} \right\|_{E_n^\theta} \leq \frac{\varepsilon_0}{16 M_2 C^*}, \quad \|v^{n,\pm}\|_{E^\beta} \leq \hat{C} \rho_0$$
$$(4.3.18)$$

(see (4.2.14) and (4.2.18)). At the next step, we apply Proposition 4.1 with boundary values $v^{n,\pm}$ and β instead of θ. Choose $\hat{\rho}_2$ such that (cf. (4.3.14) and (4.3.3))

$$\tilde{C}_\theta C_{[\beta-\theta]}\hat{\rho}_2 \leq \tilde{\rho}_2,$$

require $(C_{[\gamma-\theta]} + \hat{C})\rho_0 \leq \tilde{\rho}_2$, and denote the unique solution in $C([0,T];\mathcal{U}_{E^\beta}(0;\hat{\rho}_2))$ by $v^n(t)$, $0 \leq t \leq T$. We will show that

$$u^{n,0} = v^n(0) + u^*, \quad u(t;u^{n,0}) = v^n(t) + u^*, \quad 0 \leq t \leq T, \qquad (4.3.19)$$

satisfies (4.3.7). For this purpose, we substitute $v_n(\cdot)$ and $w_n(\cdot) := \tilde{p}_n^\theta v^n(\cdot)$ into (4.2.18). This inequality is valid since

$$\|v_n(t)\|_{E_n^\theta} \leq C_{[\gamma-\theta]}\rho_0,$$

$$\|w_n(t)\|_{E_n^\theta} \leq \tilde{C}_\theta\|v^n(t)\|_{E^\theta} \leq \tilde{C}_\theta C_{[\beta-\theta]}\|v^n(t)\|_{E^\beta} \leq \tilde{C}_\theta C_{[\beta-\theta]}\hat{\rho}_2 \leq \tilde{\rho}_2,$$

$$\|v_n^\pm\|_{E_n^\theta} \leq C_{[\gamma-\theta]}\|v_n^\pm\|_{E_n^\gamma} \leq C_{[\gamma-\theta]}C_b\rho_0 \leq \tilde{\rho}_1.$$

We obtain the estimate

$$\|v_n - w_n\|_{C\left([0,T];E_n^\theta\right)} \leq C^*\left\|w_n - G_n(v_n^-, v_n^+, w_n)\right\|_{C\left([0,T];E_n^\theta\right)}$$

$$\leq C^* \sup_{0 \leq t \leq T} \left\|\eta_n^-(t) + \eta_n^+(t) + \varphi_n^-(t) + \varphi_n^+(t)\right\|_{E_n^\theta},$$

$$\qquad (4.3.20)$$

where the terms on the right-hand side are given by the formulas

$$\eta_n^-(t) = \tilde{p}_n^\theta \exp\left(tA_{u^*}\right)(I-P)v^{n,-} - \exp(tA_{u_n^*,n})(I_n - P_n)\tilde{p}_n^\theta v^{n,-}$$

$$+ \tilde{p}_n^\theta \int_0^t \exp\left((t-s)A_{u^*}\right)(I-P)F_{u^*}(v^n(s))ds$$

$$- \int_0^t \exp\left((t-s)A_{u_n^*,n}\right)(I_n - P_n)F_{u_n^*,n}\left(\tilde{p}_n^\theta v^n(s)\right)ds,$$

$$\eta_n^+(t) = \tilde{p}_n^\theta \exp\left((t-T)A_{u^*}\right)Pv^{n,+} - \exp\left((t-T)A_{u_n^*,n}\right)P_n\tilde{p}_n^\theta v^{n,+}$$

$$+ \tilde{p}_n^\theta \int_t^T \exp\left((t-s)A_{u^*}\right)PF_{u^*}(v^n(s))ds$$

$$- \int_t^T \exp\left((t-s)A_{u_n^*,n}\right)P_nF_{u_n^*,n}\left(\tilde{p}_n^\theta v^n(s)\right)ds,$$

$$\varphi_n^-(t) = \exp\left(tA_{u_n^*,n}\right)(I_n - P_n)\left(\tilde{p}_n^\theta v^{n,-} - v_n^-\right),$$

$$\varphi_n^+(t) = \exp\left((t-T)A_{u_n^*,n}\right)P_n\left(\tilde{p}_n^\theta v^{n,+} - v_n^+\right).$$

Using Theorem 4.1, we obtain the estimate

$$\|\eta_n^\pm\| \leq \frac{\varepsilon_0}{8C^*}$$

similarly to the estimate (4.3.12) in Theorem 4.2; the main difference is that $v(s)$ is replaced by $v^n(s)$ and the compact sets are given by

$$K_2 = \left\{(I - P)F_{u^*}(w) : \ w \in \mathcal{U}_{E^\beta}(0; \hat{\rho}_2)\right\}, \quad K_1 = \mathcal{U}_{E^\beta}(0; \hat{C}\rho_0)$$

(cf. (4.3.18)).

For the second term, we have from Theorem 3.16 and (4.3.18)

$$\left\|\varphi_n^-(t) + \varphi_n^+(t)\right\|_{E_n^\theta} \leq M_2 e^{-\tilde{\beta}t}\frac{\varepsilon_0}{8M_2C^*} \leq \frac{\varepsilon_0}{8C^*}.$$

Using this inequality, the estimate (4.3.20), and Proposition 3.5, for large n we finally have

$$\left\|u_n(t) - p_n^\theta u(t, u^{n,0})\right\|_{E_n^\theta} \leq \left\|v_n(t) - p_n^\theta v^n(t)\right\|_{E_n^\theta} + \left\|u_n^* - p_n^\theta u^*\right\|_{E_n^\theta} \leq \varepsilon_0.$$

The theorem is proved.   □

4.4   Shadowing in Time

Due to the splitting of the spaces, in addition to p_n and p_n^θ, we need other connecting maps. We also introduce the operators

$$\hat{p}_n = P_n(\sigma_n^+)\tilde{p}_n^\theta P(\sigma^+), \quad \check{p}_n = (I_n - P_n(\sigma_n^+))\tilde{p}_n^\theta(I - P(\sigma^+)).$$

Since the operators $P_n(\sigma_n^+)$ and p_n are uniformly bounded, the operators $\check{p}_n$ and $\hat{p}_n$ are also uniformly bounded on E^θ.

We reformulate Theorems 4.2 and 4.3 as follows.

Theorem 4.4 (see [56]). *Let the operator A be the generator of exponentially decreasing, analytic C_0-semigroup and u^* be a hyperbolic equilibrium point of the problem (4.1.5). Let $\Delta_{cc} \neq \varnothing$,*

$$f_n(x_n) \xrightarrow{\ \mathcal{P}\ } f(x), \quad f_n'(x_n) \xrightarrow{\ \mathcal{P}^\theta\mathcal{P}\ } f'(x) \quad as \ x_n \xrightarrow{\ \mathcal{P}^\theta\ } x,$$

and let Conditions (F1)–(F3) be fulfilled. Then for any mild solution of the problem (4.1.9) with $v^0, v^T \in \mathcal{U}_{E^\gamma}(0; \rho)$, there exist mild solutions of the problem (4.2.15) with $v_n^0 = \check{p}_n v^0$ and $v_n^T = \hat{p}_n v^T$ such that

$$\sup_{v^0, v^T \in \mathcal{U}_{E^\gamma}(0;\rho)} \sup_{0 \leq t \leq T} \left\|v_n(t; v_n^0, v_n^T) - \tilde{p}_n^\theta v(t; v^0, v^T)\right\|_{E_n^\theta} \to 0 \qquad (4.4.1)$$

as $n \to \infty$.

Theorem 4.5 (see [56]). *Let the operator A be the generator of exponentially decreasing, analytic C_0-semigroup and u^* be a hyperbolic equilibrium point of the problem (4.1.5). Let $\Delta_{cc} \neq \varnothing$,*

$$f_n(x_n) \xrightarrow{\mathcal{P}} f(x), \quad f_n'(x_n) \xrightarrow{\mathcal{P}^\theta \mathcal{P}} f'(x) \quad as \ x_n \xrightarrow{\mathcal{P}^\theta} x,$$

and let Conditions (F1) and (F3) be fulfilled. Then for any sequence of mild solutions of the problems (4.2.15) with $v_n^0, v_n^T \in \mathcal{U}_{E_n^\gamma}(0; \rho)$, $0 \leq \theta < \gamma < 1$, there are sets of elements $\{v^{0,E}\}$, $v^{0,E} \in E^\theta$, and $\{v^{T,E}\}$, $v^{T,E} \in E^\theta$, such that $v^{0,E}, v^{T,E} \in \mathcal{U}_{E^\theta}(0; \rho)$, $0 \leq \theta < \gamma < 1$, and mild solutions of the problems (4.1.9) with $v^{0,E}$, $v^{T,E} \in \mathcal{U}_{E^\theta}(0; \rho)$ such that

$$\sup_{v_n^0, v_n^T \in \mathcal{U}_{E^\gamma}(0;\rho)} \sup_{0 \leq t \leq T} \left\| v_n(t; v_n^0, v_n^T) - \bar{p}_n^\theta v(t; v^{0,E}, v^{T,E}) \right\|_{E_n^\theta} \to 0 \quad (4.4.2)$$

as $n \to \infty$.

4.4.1   Shadowing in time and the condition $\Delta_{cc} \neq \varnothing$

For the problem (4.2.11), let us consider the time discretization scheme

$$\frac{V_n(t + \tau_n) - V_n(t)}{\tau_n} = A_{u_n^*, n} V_n(t + \tau_n) + F_{u_n^*, n}(V_n(t)), \quad t = k\tau_n, \quad (4.4.3)$$

with the initial data $V_n(0) = v_n^0$. The solution of such problem is given by the formula

$$\begin{aligned}
V_n(t + \tau_n) &= \left(I_n - \tau_n A_{u_n^*, n} \right)^{-1} V_n(t) \\
&\quad + \tau_n \left(I_n - \tau_n A_{u_n^*, n} \right)^{-1} F_{u_n^*, n}(V_n(t)) \\
&= \left(I_n - \tau_n A_{u_n^*, n} \right)^{-k} V_n(0) \\
&\quad + \tau_n \sum_{j=0}^{k} \left(I_n - \tau_n A_{u_n^*, n} \right)^{-(k-j+1)} F_{u_n^*, n}\left(V_n(j\tau_n) \right), \quad (4.4.4)
\end{aligned}$$

where $t = k\tau_n$ and $V_n(0) = v_n^0$.

Theorem 4.6 (see [223]). *Let $\Delta_{cc} \neq \varnothing$ and let the resolvents of the operators A_n and A be compact operators. Assume also that the C_0-semigroup $e^{tA_{u^*}}$, $t \in \overline{\mathbb{R}}^+$, is hyperbolic and Condition (B_1) is satisfied. Then*

$$\begin{cases}
\left\| \left(I_n - \tau_n A_{u_n^*, n} \right)^{-k_n} (I_n - P_n) \right\|_{E_n} \leq M_2 r^{[t]}, & t = k_n \tau_n \geq 0, \\
\left\| \left(I_n - \tau_n A_{u_n^*, n} \right)^{k_n} P_n \right\|_{E_n} \leq M_2 r^{-[t]}, & t = -k_n \tau_n \leq 0,
\end{cases} \quad (4.4.5)$$

where $r < 1$.

Proof. The compact convergence of resolvents

$$(\lambda I_n - A_n)^{-1} \xrightarrow{\mathcal{PP}} (\lambda I - A)^{-1}$$

implies (see [69]) that

$$\left(\lambda I_n - A_{u_n^*,n}\right)^{-1} \xrightarrow{\mathcal{PP}} \left(\lambda I - A_{u^*}\right)^{-1} \quad \text{compactly.}$$

We set $B_n = (I_n - \tau_n A_{u_n^*,n})^{-k_n}$, $\tau_n k_n = 1$, and $B = e^{1 A_{u^*}}$. We conclude that $B_n \xrightarrow{\mathcal{PP}} B$ since the operators $A_{u_n^*,n}$ and A_{u^*} are consistent. Note that by Condition (B$_1$) one has (see [120, 267])

$$\left\| \tau_n k_n A_{u_n^*,n} \left(I_n - \tau_n A_{u_n^*,n} \right)^{-k_n} \right\|_{B(E_n)} \leq \text{const,}$$

which implies that

$$B_n = A_{u_n^*,n}^{-1} \tau_n k_n A_{u_n^*,n} \left(I_n - \tau_n A_{u_n^*,n} \right)^{-k_n} \xrightarrow{\mathcal{PP}} B \quad \text{compactly}$$

since $A_{u_n^*,n}^{-1} \xrightarrow{\mathcal{PP}} A_{u^*}^{-1}$ compactly. Now one gets discrete dichotomy for B_n by Theorem 4.13. Theorem 4.6 is proved. $\qquad\square$

Remark 4.3. One can consider the problem (4.4.3) involving the term $F_{u_n^*,n}(V_n(t + \tau_n))$, but it does not make much of the theoretical value.

The problem (4.2.15) can also be discretized following the approach (4.4.3)–(4.4.4), so we have

$$\frac{V_n(t + \tau_n) - V_n(t)}{\tau_n} = A_{u_n^*,n} V_n(t + \tau_n) + F_{u_n^*,n}(V_n(t)), \quad t = k\tau_n, \tag{4.4.6}$$

$$(I_n - P_n)V_n(0) = (I_n - P_n)v_n^0, \quad P_n V_n(T) = P_n v_n^T.$$

A solution of the problem (4.4.6) can be obtained by the formulas

$$(I_n - P_n)V_n(t + \tau_n) = \left(I_n - \tau_n A_{u_n^*,n} \right)^{-1} (I_n - P_n)V_n(t)$$
$$+ \tau_n \left(I_n - \tau_n A_{u_n^*,n} \right)^{-1} (I_n - P_n) F_{u_n^*,n}\left(V_n(j\tau_n) \right)$$

and

$$\left(I_n - \tau_n A_{u_n^*,n} \right) P_n V_n(t + \tau_n) = P_n V_n(t) + \tau_n P_n F_{u_n^*,n}\left(V_n(k\tau_n) \right), \quad t = k\tau_n.$$

The last equation is solvable since $F_{u_n^*,n}(V_n(k\tau_n))$ is sufficiently small in the neighborhood considered (see Condition (F3)). Thus, we have the following representation of the solution of the problem (4.4.6):

$$V_n\left(t; v_n^0, v_n^T\right) = (I_n - P_n)V_n(t) + P_n V_n(t)$$
$$= \left(I_n - \tau_n A_{u_n^*,n} \right)^{-k+1} (I_n - P_n)v_n^0$$

$$+ \tau_n \sum_{j=0}^{k-1} \left(I_n - \tau_n A_{u_n^*,n}\right)^{-(k-j)} (I_n - P_n) F_{u_n^*,n}\left(V_n(j\tau_n)\right)$$

$$+ \left(I_n - \tau_n A_{u_n^*,n}\right)^{K-k} P_n v_n^T$$

$$- \tau_n \sum_{j=k}^{K-1} \left(I_n - \tau_n A_{u_n^*,n}\right)^{j-k} P_n F_{u_n^*,n}\left(V_n(j\tau_n)\right), \quad t = k\tau_n.$$

Lemma 4.4. *Let the operator A be the generator of an exponentially decreasing, analytic C_0-semigroup and u^* be a hyperbolic equilibrium point of the problem (4.1.5). Let $\Delta_{cc} \neq \varnothing$,*

$$f_n(x_n) \xrightarrow{\mathcal{P}} f(x), \quad f_n'(x_n) \xrightarrow{\mathcal{P}^\theta \mathcal{P}} f'(x) \quad as \; x_n \xrightarrow{\mathcal{P}^\theta} x,$$

and let Conditions (F1)–(F3) and (B$_1$) be fulfilled. Then there is a constant $C > 0$ independent of n such that

$$\left\| \left(I_n - \tau_n A_{u_n^*,n}\right)^{-k} (I_n - P_n) \right\| \leq C,$$

$$\tau_n \sum_{j=0}^{k-1} \left\| (-A_n)^\theta \left(I_n - \tau_n A_{u_n^*,n}\right)^{-(k-j)} (I_n - P_n) \right\| \leq C$$

for $k, n \in \mathbb{N}$.

Proof. By Theorem 4.2, the operators $A_{u_n^*,n}$ generate on the subspaces $(I_n - P_n) E_n$ bounded analytic C_0-semigroups with the uniform estimates

$$\left\| e^{t A_{u_n^*,n}} (I_n - P_n) \right\| \leq M e^{-\omega t}, \quad t \geq 0.$$

This implies

$$\left\| (I_n - \tau_n A_{u_n^*,n})^{-k} (I_n - P_n) \right\| \leq C$$

for any $k, n \in \mathbb{N}$ and $0 < \tau_n \leq \tau^*$. Now using Theorem 4.6 we get the dichotomy estimates (4.4.5) and the following formula with an absolutely converging integral:

$$\tau_n \sum_{j=0}^{k-1} (-A_n)^\theta \left(I_n - \tau_n A_{u_n^*,n}\right)^{-(k-j)} (I_n - P_n)$$

$$= (-A_n)^\theta (-A_{u_n^*,n})^{-\theta} \tau_n \sum_{j=0}^{k-1} \frac{1}{2\pi i}$$

$$\times \int_{\Gamma_1} \zeta^\theta (1 - \tau_n \zeta)^{-(k-j)} \left(\zeta I_n - A_{u_n^*,n}\right)^{-1} d\zeta (I_n - P_n), \qquad (4.4.7)$$

where the contour Γ_1 corresponds to the spectrum $\sigma(A_{u_n^*,n}(I_n - P_n))$. It is also clear that for $\zeta \in \Gamma_1$ one has

$$|1 - \tau_n\zeta| = \sqrt{(1 - \tau_n \operatorname{Re}\zeta)^2 + (\tau_n \operatorname{Im}\zeta)^2} \geq 1 - \tau_n \operatorname{Re}\zeta.$$

Therefore,

$$|1 - \tau_n\zeta|^{-1} \leq \frac{1}{1 - \tau_n \operatorname{Re}\zeta},$$

$$\sum_{j=0}^{k-1} |1 - \tau_n\zeta|^{-(k-j)} \leq \frac{1}{|1 - \tau_n\zeta|} \frac{1 - \tau_n \operatorname{Re}\zeta}{-\tau_n \operatorname{Re}\zeta}, \quad k, n \in \mathbb{N},$$

for $\zeta \in \Gamma_1$. The integral in (4.4.7) absolutely converges. Indeed,

$$\int_0^q \tau_n \sum_{j=0}^{k-1} |\zeta|^\theta |1 - \tau_n\zeta|^{-(k-j)} \frac{M}{|\zeta| + 1} d|\zeta|$$

$$+ \int_q^\infty \tau_n \sum_{j=0}^{k-1} \zeta^\theta (1 - \tau_n\zeta)^{-(k-j)} \frac{M}{|\zeta| + 1} d|\zeta|$$

$$\leq \int_0^q |\zeta|^\theta \frac{M}{|\zeta| + 1} d|\zeta| \frac{\tau_n}{1 - \frac{1}{1 + c\tau_n}}$$

$$+ \int_q^\infty |\zeta|^\theta \frac{1}{1 - \frac{1}{1 - \tau_n \operatorname{Re}\zeta}} \frac{M}{|\zeta| + 1} \frac{d|\zeta|}{|1 - \tau_n\zeta|}$$

$$\leq \frac{(1 + c\tau_n)}{c} \int_0^q |\zeta|^\theta \frac{M}{|\zeta| + 1} d\zeta$$

$$+ \int_q^\infty |\zeta|^\theta \frac{(1 - \tau_n \operatorname{Re}\zeta) d|\zeta|}{-\operatorname{Re}\zeta |1 - \tau_n\zeta|(|\zeta| + 1)}.$$

From [56] we have

$$\left\| (A_{u_n^*,n})^\theta (A_n)^{-\theta} \right\|, \quad \left\| (A_n)^\theta (A_{u_n^*,n})^{-\theta} \right\| \leq C, \quad n \in \mathbb{N}.$$

The second statement is proved. $\square$

Lemma 4.5. *Let the operator A be the generator of exponentially decreasing, analytic C_0-semigroup and u^* be a hyperbolic equilibrium point of the problem (4.1.5). Let $\Delta_{cc} \neq \varnothing$,*

$$f_n(x_n) \xrightarrow{\mathcal{P}} f(x), \quad f_n'(x_n) \xrightarrow{\mathcal{P}^\theta \mathcal{P}} f'(x) \quad \text{as } x_n \xrightarrow{\mathcal{P}^\theta} x,$$

and let Conditions (F1)–(F3) and (B$_1$) be fulfilled. Then there is a constant $C > 0$ independent of n such that

$$\left\| (I_n - \tau_n A_{u_n^*,n})^k P_n \right\| \leq C,$$

$$\tau_n \sum_{j=k}^{K-1} \left\| (-A_n)^\theta (I_n - \tau_n A_{u_n^*,n})^{j-k} P_n \right\| \le C$$

for $K, k, n \in \mathbb{N}$, $k \le K - 1$.

Proof. Theorem 4.6 implies the existence of projectors P_n such that $P_n \xrightarrow{\;PP\;} P$ compactly. Using the formulas

$$(-A_n)^\theta (I_n - \tau_n A_{u_n^*,n})^j P_n$$

$$= (-A_n)^\theta (A_{u_n^*,n})^{-\theta} \frac{1}{2\pi i} \int_{\Gamma_2} \zeta^\theta (1 - \tau_n \zeta)^j (\zeta I_n - A_{u_n^*,n})^{-1} d\zeta P_n,$$

$$(-A_n)^\theta \tau_n \sum_{j=0}^{K-k-1} (I_n - \tau_n A_{u_n^*,n})^j P_n$$

$$= (-A_n)^\theta (A_{u_n^*,n})^{-\theta} \frac{1}{2\pi i} \int_{\Gamma_2} \tau_n$$

$$\times \sum_{j=0}^{K-k-1} \zeta^\theta (1 - \tau_n \zeta)^j (\zeta I_n - A_{u_n^*,n})^{-1} d\zeta P_n,$$

we get estimates, where Γ_2 is a finite contour corresponding to the spectrum $\sigma(A_{u_n^*,n} P_n)$. Indeed, for any $\zeta \in \Gamma_2$, we have

$$|1 - \tau_n \zeta| \le 1 - 2\tau_n \operatorname{Re} \zeta + (\tau_n \operatorname{Re} \zeta)^2 + (\tau_n \operatorname{Im} \zeta)^2 \le 1 - c\tau_n < 1, \quad \zeta \in \Gamma_2,$$

with some constant $c > 0$ and sufficiently small τ_n since Γ surrounds a finite part of the spectrum $\sigma(A_{u_n^*,n} P_n)$ on the right half-plane. Thus,

$$\tau_n \sum_{j=0}^{K-k-1} |1 - \tau_n \zeta|^j \le \tau_n \frac{1}{1 - (1 - c\tau_n)} \le \frac{1}{c}, \quad K, k, n \in \mathbb{N}, \quad k \le K - 1,$$

which implies the lemma. $\qquad \square$

Lemma 4.6. *Let the operators A_n be generators of, uniformly in n, exponentially decreasing analytic C_0-semigroup, i.e.,*

$$\|e^{tA_n}\| \le M e^{-\omega t}, \quad t \ge 0.$$

Then

$$(I_n - \tau_n A_n)^{-k} \to 0, \quad (-A_n)^\theta (I_n - \tau_n A_n)^{-k} \to 0, \quad 0 < \theta < 1,$$

for $0 < \tau_n \le \tau^*$, $k \to \infty$, $k\tau_n \to \infty$, and

$$\left\| \left((I_n - \tau_n A_n)^{-k} - e^{tA_n} \right)(-A_n)^{-\theta} u_n^0 \right\| \le C \frac{\tau_n^\theta}{k} \|u_n^0\|. \tag{4.4.8}$$

Proof. For $t = k\tau_n$, consider the difference

$$
\left(I_n - \tau_n A_n\right)^{-k} - e^{tA_n}
$$

$$
= \int_0^{\tau_n} \frac{d}{ds}\left(\exp\left(k(\tau_n - s)A\right)(I_n - sA_n)^{-k}\right) ds
$$

$$
= \int_0^{\tau_n} \exp\left(k(\tau_n - s)A\right) k \left(-A_n + \frac{A_n}{I_n - sA_n}\right) (I_n - sA_n)^{-k} ds
$$

$$
= \frac{1}{k} \int_0^{\tau_n} \frac{\left((\tau_n - s)k(-A_n)\right)^{1/2} \exp\left(k(\tau_n - s)A\right)}{\sqrt{\tau_n - s}}
$$

$$
\times \frac{\left(sk(-A_n)\right)^{3/2}(I_n - sA_n)^{-k-1}}{\sqrt{s}} ds. \qquad (4.4.9)
$$

It is known (see [235]) that

$$
\left\| (t(-A_n))^{1/2} e^{tA_n} \right\| \le M e^{-\omega t}, \quad t \ge 0.
$$

We can see that

$$
\left\| (ks(-A_n))^{3/2}(I_n - sA_n)^{-k} \right\| \le C
$$

for any $k, n \in \mathbb{N}$, $k > \theta$, and $0 < s \le \tau^*$. Indeed,

$$
\left\| (-A_n)^\theta \left(\lambda I_n - A_n\right)^{-k} \right\| \le \frac{1}{(k-1)!} \int_0^\infty t^{k-1} e^{-\lambda t} \left\| (-A_n)^\theta e^{tA_n} \right\| dt
$$

$$
\le \frac{M \lambda^{\theta - k}}{(k-1)!} \int_0^\infty s^{k-1-\theta} e^{-s} ds,
$$

$$
\left\| (-A_n)^\theta (I_n - \tau_n A_n)^{-k} \right\| = \tau_n^{-k} \left\| (-A_n)^\theta \left(\frac{1}{\tau_n} I_n - A_n\right)^{-k} \right\|
$$

$$
\le M \tau_n^{-\theta} \frac{\Gamma(k - \theta)}{\Gamma(k)}.
$$

Since

$$
\lim_{k \to \infty} \frac{k^\theta \Gamma(k)}{\Gamma(k + \theta)} = 1,
$$

we get the required estimate. Now from (4.4.9) we have

$$
\left\| (I_n - \tau_n A_n)^{-k} \right\| \le \frac{1}{k}, \quad t > 0.
$$

Thus, for smooth elements like $(-A_n)^{-\theta} u_n^0$ we get (4.4.8). $\qquad \square$

Theorem 4.7 (see [223]). *Let the operator A be the generator of exponentially decreasing, analytic C_0-semigroup and u^* be a hyperbolic equilibrium point of the problem (4.1.5). Let $\Delta_{cc} \neq \varnothing$,*

$$f_n(x_n) \xrightarrow{\mathcal{P}} f(x), \quad f'_n(x_n) \xrightarrow{\mathcal{P}^\theta \mathcal{P}} f'(x) \quad as \ x_n \xrightarrow{\mathcal{P}^\theta} x,$$

and let Conditions (B_1) *and* $(F1)$–$(F3)$ *be fulfilled. Then there exists $\rho_0 > 0$ with the following property: for any ϵ_0 and any mild solution $v(\cdot)$ of the problem (4.1.9) with $v^0, v^T \in \mathcal{U}_{E^\gamma}(0; \rho_0)$, $0 \leq \theta < \gamma < 1$, there are a number $n(\epsilon_0)$ and solutions $V_n(\cdot)$ of the problems (4.4.6) with $v_n^0, v_n^T \in \mathcal{U}_{E_n^\theta}(0; \rho_0)$ such that*

$$\sup_{0 \leq t \leq T} \left\| V_n(t; v_n^0, v_n^T) - \tilde{p}_n^\theta v(t; v^0, v^T) \right\|_{E_n^\theta} \leq \epsilon_0 \quad for \ n \geq n(\epsilon_0). \quad (4.4.10)$$

Proof. As we have already mentioned, a solution of the problem (4.4.6) exists; this can be proved in the same way as [56, Proposition 2.2]. Indeed, for $K\tau_n = T$, we introduce (as in [120, Sec. 5.1]) the Banach space $X_n = C_{\tau_n}([0, T]; E_n^\theta)$ with the norm

$$\|V_n(\cdot)\|_{C_{\tau_n}([0,T];E_n^\theta)} = \max_{0 \leq j \leq K} \left\| V_n(j\tau_n) \right\|_{E_n^\theta}.$$

We denote the operator on the right-hand side of (4.4.4) applied to an element $V_n(.) \in C_{\tau_n}([0, T]; E_n^\theta)$ by $G(v_n^0, v_n^T; V_n(\cdot))$. For such an operator, one has $G'(v_n^0, v_n^T; 0) = 0$. Therefore, the operator $I_n - G'(v_n^0, v_n^T; 0)$ is invertible. Moreover (see [56, (2.17)]), for the Fréchet derivative we have

$$\left\| G'(v_n^0, v_n^T; 0) - G'(v_n^0, v_n^T; V_n(\cdot)) \right\|_{B(X_n)} \leq q < 1,$$

if $V_n(\cdot) \in \mathcal{U}_{C_{\tau_n}([0,T];E_n^\theta)}(0; \rho_0)$ since the series converge according to Lemmas 4.4 and 4.5. Thus, the existence of a unique solution $V_n(\cdot)$ of the problem (4.4.4) follows from [288, Lemma 1].

Now we prove that if $v_n(\cdot)$ is a solution of (4.2.15), then

$$\sup_{0 \leq t \leq T} \left\| v_n(t; v_n^0, v_n^T) - V_n(t) \right\|_{E_n^\theta} \leq \frac{\epsilon_0}{3}$$

for $n \geq n(\epsilon_0)$; this follows from (4.4.8) with $\epsilon_0/3$ on the right-hand side. We recall here that

$$v_n(t) = u_n(t) - u_n^*, \quad v(t) = u(t) - u^*.$$

by the uniqueness of solutions. Since

$$v_n(t) - \tilde{p}_n^\theta v(t) = u_n(t) - \tilde{p}_n^\theta u(t) - u_n^* + \tilde{p}_n^\theta u^*,$$

$$\left\| u_n^* - \tilde{p}_n^\theta u^* \right\| \leq \frac{\epsilon_0}{3} \quad for \ n \geq n(\epsilon_0),$$

the inequality (4.4.8) can be used for estimating $v_n(t) - \tilde{p}_n^\theta v(t)$. To estimate $\|v_n(t) - V_n(k\tau_n)\|$, we consider the difference of the right-hand sides of (4.4) and (4.2.16). For example, for v_n^0 and $\gamma - \theta > 0$ we have (see [231])

$$\left\| (-A_n)^\theta \left(\exp(tA_{u_n^*,n}) - \left(I_n - \tau_n A_{u_n^*,n} \right)^{-k} \right) (I_n - P_n) v_n^0 \right\|$$

$$\leq C \left\| (-A_{u_n^*,n})^\theta \left(\exp(tA_{u_n^*,n}) - \left(I_n - \tau_n A_{u_n^*,n} \right)^{-k} \right) (I_n - P_n) v_n^0 \right\|$$

$$\leq C \frac{\tau_n^{\gamma-\theta}}{k} \left\| (-A_{u_n^*,n})^\gamma v_n^0 \right\|, \tag{4.4.11}$$

where the first inequality follows from [56, Remark 4.2] and the second inequality coincides with (4.4.8). The estimates of the integral terms follow from Condition (F3) (see Remark 4.6) and Lemmas 4.4–4.5. Now we can choose τ_n so that (4.4.10) holds. $\square$

Theorem 4.8. *Let the operator A be the generator of exponentially decreasing, analytic C_0-semigroup and u^* be a hyperbolic equilibrium point of the problem (4.1.5). Let $\Delta_{cc} \neq \varnothing$,*

$$f_n(x_n) \xrightarrow{\;\mathcal{P}\;} f(x), \quad f_n'(x_n) \xrightarrow{\;\mathcal{P}^\theta \mathcal{P}\;} f'(x) \quad \text{as } x_n \xrightarrow{\;\mathcal{P}^\theta\;} x,$$

and let Conditions (B_n) and $(F1)$–$(F3)$ be fulfilled. Then there exists $\rho_0 > 0$ with the following property: for any $\epsilon_0 > 0$, there exists $n_0 = n(\epsilon_0) \in \mathbb{N}$ such that for any solution $V_n(t)$, $n \geq n_0$ of (4.4.6) satisfying the condition $V_n(t) \in \mathcal{U}_{E_n^\gamma}(0; \rho_0)$, $0 \leq t \leq T$, $0 \leq \theta < \gamma < 1$, for some $0 < T \leq \infty$ with $v_n^0, v_n^T \in \mathcal{U}_{E_n^\theta}(0; \rho)$, there exist elements $v^{n,0}, v^{n,T} \in E^\theta$, $n \geq n_0$, such that a mild solution $v(t; v^{n,0}, v^{n,T})$ of the problem (4.1.9) exists on $[0, T]$ and satisfies the estimate

$$\sup_{0 \leq t \leq T} \left\| V_n(t; v_n^0, v_n^T) - \tilde{p}_n^\theta v(t; v^{n,0}, v^{n,T}) \right\|_{E_n^\theta} \leq \epsilon_0 \quad \forall n \geq n(\epsilon_0). \tag{4.4.12}$$

The proof is based on Theorem 4.5 and estimates similar to (4.4.11).

Remark 4.4. One can prove theorems similar to Theorems 4.7 and 4.8 for the scheme

$$\frac{V_n(t + \tau_n) - V_n(t)}{\tau_n} = A_{u_n^*,n} V_n(t) + F_{u_n^*,n}(V_n(t)), \quad t = k\tau_n,$$

instead of the scheme (4.4.6), but under some stability condition like $\left\| \tau_n A_{u_n^*} \right\| \leq \text{const.}$

4.5 Checking of the Condition $\Delta_{cc} \neq \varnothing$

In this section, we show how the conditions of Theorems 4.2 and 4.3 can be satisfied for the finite-element and finite-difference methods.

Example 4.1. Let $\Omega \subset \mathbb{R}^d$ be a bounded smooth domain. Consider the second-order strongly elliptic operator

$$Lu(x) = \sum_{i,j=1}^{d} a_{ij}(x) u_{x_i x_j}(x) + \sum_{j=1}^{d} b_j(x) u_{x_j}(x) + c(x)u(x), \qquad (4.5.1)$$

where the coefficients a_{ij}, b_j, and c are smooth bounded functions. Consider the associated parabolic problem

$$\begin{aligned}
u_t(t,x) &= Lu(t,x) + f(u(t,x)), & t > 0, \quad x \in \Omega, \\
u(t,x) &= 0, & t > 0, \quad x \in \partial\Omega, \qquad (4.5.2) \\
u(0,x) &= u^0(x) \in H_0^1(\Omega).
\end{aligned}$$

Let $E = L^2(\Omega)$. We define the operator $A : D(A) \subset E \to E$ such that $D(A) = H^2(\Omega) \cap H_0^1(\Omega)$ and $Au = Lu$ for all $u \in D(A)$. It is well known that A generates an analytic and compact C_0-semigroup $\{\exp(tA) : t \geq 0\}$. Assume that $c(x)$ is such that the spectrum of A is located to the left of the imaginary axis. Then, we can define the fractional powers $(-A)^\theta$ of the operator $-A$ as before. It is well known that $E^1 = D(A) = H^2(\Omega) \cap H_0^1(\Omega)$ and $E^{1/2} = H_0^1(\Omega)$.

As to the nonlinear term $f(\cdot)$, it is known (see [22,173]) that under some growth conditions, the problem (4.5.2) is locally well-posed in $E^{1/2}$ and the operator-function $f(\cdot)$ is Fréchet differentiable as a function from E^θ to E. For example, these assumptions are as follows (cf. [173]): the scalar function $f(x, \cdot) : \mathbb{R} \to \mathbb{R}$, $x \in \Omega$, belongs to the class $C^2(\mathbb{R}, \mathbb{R})$ and

$$\left| f_\xi^{(l)}(x, \xi) \right| \leq C\left(1 + |\xi|^{\delta + 1 - l}\right),$$

for any $\xi \in \mathbb{R}$ and $x \in \Omega$, where $l = 1, 2$, and $\delta = 2$ for $d = 3$ and $\delta \in [1, \infty)$ for $d = 2$. Then one can show (see [173]) that

$$\left\| f'(u) - f'(v) \right\|_{B(E^{1/2}, E)} \leq C(\rho) \|u - v\|_{E^{1/2}}, \qquad (4.5.3)$$

$$\left\| f(u) - f(v) - f'(w)(u - v) \right\|_{E^{(k-1)/2}} $$
$$\leq C(\rho)\left(\|u - w\|_{E^{1/2}} + \|v - w\|_{E^{1/2}} \right) \|u - v\|_{E^{k/2}} \qquad (4.5.4)$$

for $k = 1, 2$ and any $v, w, u \in \{z \in E : \|z - u^*\|_{E^{1/2}} \leq \rho\}$. The inequalities (4.5.3)–(4.5.4) imply Condition (F1).

The problem (4.1.3) with the operator A and the function $f(\cdot)$ defined as above is well posed and possesses all properties needed for our main theorems.

Moreover, the approximation problems (4.2.9) also have the required properties when defined by the finite-element method.

Indeed, assume that A, E, and $E^{1/2}$ are as before. It is well known (see [146]) that A is in one-to-one correspondence with a sesquilinear form $a : E^{1/2} \times E^{1/2} \to \mathbb{C}$ such that

$$
\begin{aligned}
|a(u,v)| &\leq c_1 \|u\|_{E^{1/2}} \|v\|_{E^{1/2}}, &\quad& u, v \in E^{1/2}, \\
\operatorname{Re} a(u,u) &\geq c_2 \|u\|_{E^{1/2}}, &\quad& u \in E^{1/2}, \\
a(u,v) &= \langle -Au, v \rangle, &\quad& u \in D(A), \; v \in E^{1/2}.
\end{aligned}
$$

Consider a convex polygon $\Omega \subset \mathbb{R}^2$ and a regular triangulation whose triangles have maximum diameter h. Denote by S_h the space of functions in $E^{1/2}$ that are linear in each element and vanish on the boundary. Then S_h is a family of finite dimensional subspaces of $H_0^1(\Omega)$ with the standard approximation property (see [275])

$$
\inf_{\chi \in S_h} \left(\|v - \chi\|_E + h\|v - \chi\|_{E^{1/2}} \right) \leq Ch^2 \|v\|_{E^1}, \quad v \in H^2(\Omega) \cap H_0^1(\Omega).
$$

We denote by $P_h u$ the projection of $u \in E$ onto $S_h = E_h^{1/2}$ with respect to the inner product of the space $L^2(\Omega)$. These operators play the role of the connecting mappings $\{p_h\}$. Within this framework, the finite-element approximation $A_h : S_h \to S_h$ of A is defined by the formula

$$
\langle -A_h \phi_h, \psi_h \rangle = a(\phi_h, \psi_h), \quad \phi_h, \psi_h \in E_h^{1/2}.
$$

In other words, A_h is the operator associated with the sesquilinear form $a_h(\cdot, \cdot)$, which is the restriction of $a(\cdot, \cdot)$ to $E_h^{1/2} \times E_h^{1/2}$. In this setting, we can prove (see [100]) that there exists a constant C and an acute angle θ such that for $u \in E$ and $\theta \leq |\arg z| \leq \pi$ we have

$$
\left\| (zI - A)^{-1} u - (zI_h - A_h)^{-1} P_h u \right\|_E \leq Ch^2 \|u\|_E.
$$

This estimate means $\mathcal{P}$-convergence with respect to uniform convergence of resolvents. Since our resolvent $(\lambda I - A)^{-1}$ is compact for some λ, the above inequality (where $\mu(\cdot)$ is the noncompactness measure) yields

$$
\begin{aligned}
\mu\!\left((zI_h - A_h)^{-1} x_h \right) \leq{}& \mu\!\left((zI - A)^{-1} x_h \right) \\
&+ \varlimsup_{h \to 0} \left\| (zI - A)^{-1} x_h - (zI_h - A_h)^{-1} x_h \right\|_E = 0,
\end{aligned}
$$

which implies the compact convergence of resolvents as $h \to 0$. In this way, our basic assumption $\Delta_{cc} \neq \varnothing$ can be verified.

Finally, we set $f_h(v_h) = p_h f(v_h)$ for $v_h \in E_h^{1/2}$ and

$$F_{u_h^*, h}(v_h) = f_h(v_h - u_h^*) - f_h(u_h^*) - f_h'(u_h^*)v_h;$$

then the problems (3.3.1) and (4.2.11) are well posed. The estimate (4.5.4) shows

$$\left\|F_{u^*}(v(t))\right\|_E \leq c(\rho)\|v(t)\|_{E^{1/2}}^2;$$

moreover, we have $F_{u^*}'(0) = 0$ and Condition (F3) in the form

$$\left\|F_{u_h^*, h}(v_h)\right\|_{E_n} \leq \tilde{c}(\rho)\|v_h\|_{E_h^{1/2}}^2,$$

is valid since $\|p_h\|$ is uniformly bounded.

Example 4.2. Resolvent estimates such as (B_1) were proved for finite-element and finite-difference methods, for example, in [30, 38]. Omitting details, we show how to get the compact convergence of resolvents for the finite-difference method. For example, consider in the space $E = L^2(0,1)$ the operator

$$Av(x) = \frac{d^2 v(x)}{dx^2},$$
$$D(A) = \left\{v(\cdot) \in H_0^1(0,1) \cap H^2(0,1): \; v(0) = v(1) = 0\right\}.$$

We choose the step-size $h = 1/n$ and approximate A by the operators

$$A_n u_n = \bar{\partial}_h \partial_h u_n = \left\{\frac{1}{h^2}\left(u_{n,(k+1)h} - 2u_{n,kh} + u_{n,(k-1)h}\right)\right\}_{k=1}^{n-1}, \quad (4.5.5)$$

where

$$u_{n,\cdot} \in E_n = L_h^2(0,1) = D(A_n) = \left\{\{u_{n,kh}\}_{k=1}^{n-1} \in \mathbb{R}^{n-1}\right\}.$$

Note that we set $u_{n,0} = u_{n,nh} = 0$ in (4.5.5). The connecting maps p_n are given by

$$(p_n u)_{kh} = \frac{1}{h} \int_{k-h/2}^{k+h/2)} u(x)dx, \quad u \in E \quad (4.5.6)$$

(see [285]). If the inner product is given by the formula

$$\langle u_{n,\cdot}, v_{n,\cdot}\rangle_{E_n} = h \sum_{k=1}^{n-1} u_{n,kh} v_{n,kh},$$

the formula of summation by parts yields

$$-\langle \bar{\partial}_h \, \partial_h u_n, u_n \rangle_{E_n} = \langle \partial_h u_n, \partial_h u_n \rangle_{E_n}$$

$$= \left\| \partial_h u_n \right\|_{E_n}^2 = h \sum_{k=1}^{n} \left(\frac{u_{n,kh} - u_{n,(k-1)h}}{h} \right)^2$$

(we take into account that $u_{n,0} = u_{n,nh} = 0$). This implies

$$\left\| \partial_h u_n \right\|_{E_n}^2 \leq \left\| A_n u_n \right\|_{E_n} \left\| u_n \right\|_{E_n}.$$

Hence for any bounded sequence $\{u_n\}$ such that $\{A_n u_n\}$ is also bounded in E_n, the last inequality shows that $\|u_n\|_{E_n^{1/2}}$ is bounded. This implies that $\{u_n\}$ is $\mathcal{P}$-compact. Now we can use Theorem 2.2 to get the compact convergence of resolvents.

4.6   Dichotomy for Semigroups

In this section, we consider the case where the resolvent of the operator A is not a compact operator. Thus, it is possible to examine the situation where the operator A has not only a point spectrum.

If v^0 is close to 0, i.e., $v^0 \in \mathcal{U}_{E^\theta}(0; \rho)$ with small $\rho > 0$, then a mild solution $v(t; v^0)$ of (4.1.6) can stay in the ball $\mathcal{U}_{E^\theta}(0; \rho)$ for some time. We denote the maximal time of staying $v(t; v^0)$ in $\mathcal{U}_{E^\theta}(0; \rho)$ by

$$T = T(v^0) = \sup \left\{ t \geq 0 : \; \|v(t; v^0)\|_{E^\theta} \leq \rho \text{ or } v(t; v^0) \in \mathcal{U}_{E^\theta}(0; \rho) \right\}.$$

Now, returning to a solution of (4.1.6), for any $v^0, v^T \in \mathcal{U}_{E^\theta}(0; \rho)$ we consider the boundary-value problem

$$\begin{cases} v'(t) = A_{u^*} v(t) + F_{u^*}(v(t)), & 0 \leq t \leq T, \\ (I - P(\sigma^+))v(0) = (I - P(\sigma^+))v^0, & \quad (4.6.1) \\ P(\sigma^+)v(T) = P(\sigma^+)v^T. \end{cases}$$

As is known (see [56]), a mild solution of the problem (4.6.1) satisfies the integral equation

$$v(t) = e^{(t-T)A_{u^*}} P(\sigma^+)v^T + e^{tA_{u^*}}(I - P(\sigma^+))v^0$$

$$+ \int_0^t e^{(t-s)A_{u^*}} \left(I - P(\sigma^+)\right) F_{u^*}(v(s)) ds$$

$$+ \int_t^T e^{(t-s)A_{u^*}} P(\sigma^+) F_{u^*}(v(s)) ds, \quad 0 \leq t \leq T. \quad (4.6.2)$$

For discretizing the problem (4.1.6) in space and time variables, it is very important to know what will happen to estimates of the type (3.1.10) for approximation solutions. If an estimate of the type (3.1.10) holds uniformly with respect to the parameter of discretization, then one can expect to obtain a similar behavior of approximated solutions of (4.6.2).

4.6.1 Preliminary results on dichotomy of semigroups

In this section, we consider the general approximation approach for keeping the dichotomy estimates (3.1.10) for approximations of a trajectory $u(\cdot)$ of the problem (4.1.5) in a neighborhood $\mathcal{U}_{E^\theta}(u^*; \rho)$ of an hyperbolic equilibrium point u^* by some discrete trajectories.

Let $\mathbb{T}(r) = \{\lambda : \lambda \in \mathbb{C}, |\lambda| = r\}$, $\mathbb{T} = \mathbb{T}(1)$.

Definition 4.1. A C_0-semigroup e^{tA}, $t \geq 0$, defined on a Banach space E is called *hyperbolic* if $\sigma(e^{tA}) \cap \mathbb{T} = \varnothing$ for all $t > 0$. The generator A is called *hyperbolic* if $\sigma(A) \cap i\mathbb{R} = \varnothing$.

Let us denote by $\Upsilon(\mathbb{R}; E)$ one of the spaces $L^p(\mathbb{R}; E)$, $1 \leq p < \infty$, $C_0(\mathbb{R}; E)$, or the Stepanov space $S^p(\mathbb{R}; E)$, $1 \leq p < \infty$; the space $\Upsilon(\mathbb{R}; E)$ is called the Palmer space (see [221], where the Fredholm property was first mentioned). In the Banach space $\Upsilon(\mathbb{R}; E)$, we consider the linear differential operator

$$\mathcal{L} = -\frac{d}{dt} + A : D(\mathcal{L}) \subseteq \Upsilon(\mathbb{R}; E) \to \Upsilon(\mathbb{R}; E), \qquad (4.6.3)$$

where the operator A generates a C_0-semigroup. Assume that the domain of $\mathcal{L}$ consists of functions $u(\cdot) \in \Upsilon(\mathbb{R}; E)$ such that for some function $g(\cdot) \in \Upsilon(\mathbb{R}; E)$, the relation

$$u(t) = e^{(t-s)A}u(s) - \int_s^t e^{(t-\eta)A}g(\eta)d\eta, \quad s \leq t, \quad t \in \mathbb{R},$$

holds and $\mathcal{L}u(\cdot) = g(\cdot)$. Let us note (see [44]) that the operator $\mathcal{L}$ is the generator of the C_0-semigroup $e^{t\mathcal{L}}$ on the Banach space $\Upsilon(\mathbb{R}; E)$, which is defined for all $v(\cdot) \in \Upsilon(\mathbb{R}; E)$ by the formula

$$(e^{t\mathcal{L}}v)(s) = e^{tA}v(s - t) \quad \text{for any } s \in \mathbb{R}, t \geq 0.$$

Definition 4.2. We say that a C_0-semigroup e^{tA}, $t \geq 0$, *has an exponential dichotomy on* $\mathbb{R}$ *with dichotomy data* ($M \geq 1$, $\beta > 0$) if there exists a projector $P : E \to E$ such that

(i) $e^{tA}P = Pe^{tA}$ for all $t \geq 0$;
(ii) the restriction $e^{tA}|_{\mathcal{R}(P)}$, $t \geq 0$, is invertible on $P(E)$ and

$$\left\| e^{-tA}Px \right\| \leq Me^{-\beta t}\|Px\|, \qquad\qquad t \geq 0, \ x \in E,$$
$$\left\| e^{tA}(I - P)x \right\| \leq Me^{-\beta t}\|(I - P)x\|, \quad t \geq 0, \ x \in E.$$

Theorem 4.9 (see [49]). *The operator $\mathcal{L}$ in the Banach space $\Upsilon(\mathbb{R}; E)$ is invertible if and only if the condition*

$$\sigma(e^{1A}) \cap \mathbb{T} = \varnothing \qquad (4.6.4)$$

holds. If the condition (4.6.4) is fulfilled, then

$$(\mathcal{L}^{-1}f)(t) = \int_{-\infty}^{\infty} G(t-s)f(s)ds, \quad t \in \mathbb{R}, \quad f(\cdot) \in \Upsilon(\mathbb{R}; E),$$

where the Green function has the form

$$G(\eta) = \begin{cases} -e^{\eta A} P_-, & \eta \geq 0; \\ e^{\eta A} P_+, & \eta < 0, \end{cases} \qquad (4.6.5)$$

and satisfied the estimate

$$\|G(\eta)\| \leq \begin{cases} M_+ e^{-\gamma_+ \eta}, & \eta \geq 0; \\ M_- e^{\gamma_- \eta}, & \eta < 0, \end{cases} \qquad (4.6.6)$$

where

$$M_+ = 2M\varkappa(\mathcal{L})\left(1 + \frac{1}{2\varkappa(\mathcal{L})}\right)^2, \quad M_- = 2M\varkappa(\mathcal{L})\left(1 - \frac{1}{2\varkappa(\mathcal{L})}\right)^2,$$

$$\gamma_+ = \ln\left(1 + \frac{1}{2\varkappa(\mathcal{L})}\right), \quad \gamma_- = -\ln\left(1 - \frac{1}{2\varkappa(\mathcal{L})}\right),$$

$$\varkappa(\mathcal{L}) = 1 + C(\Upsilon)\left(M + M^2\|\mathcal{L}^{-1}\|\right).$$

We note that $C(\Upsilon) = 1$ if $\Upsilon(\mathbb{R}; E) = L^\infty(\mathbb{R}, E)$ or $\Upsilon(\mathbb{R}; E) = C_0(\mathbb{R}, E)$ and $C(\Upsilon) = 2^{1-1/p}$ if $\Upsilon(\mathbb{R}; E) = L^p(\mathbb{R}, E)$ or $\Upsilon(\mathbb{R}; E) = S^p(\mathbb{R}, E)$, $p \in [1, \infty)$.

Denote by $\Upsilon(\mathbb{Z}; E)$, where $\mathbb{Z} = \{\ldots, -2, -1, 0, 1, 2, \ldots\}$, the Banach space of E-valued sequences with the corresponding discrete norm, which is consistent with the norm of $\Upsilon(\mathbb{R}; E)$. For any $u(\cdot) \in \Upsilon(\mathbb{Z}; E)$, i.e., $\{u(k)\}_{k\in\mathbb{Z}}$, and $B = e^{1A} \in B(E)$, we define the operator

$$\mathcal{B} : D(\mathcal{B}) \subseteq l^p(\mathbb{Z}; E) \to l^p(\mathbb{Z}; E),$$

$$(\mathcal{B}u)(k) = Bu(k-1), \quad k \in \mathbb{Z}, \quad u(\cdot) \in l^p(\mathbb{Z}; E).$$

Also we define the operator

$$\mathcal{D} = I - \mathcal{B} : D(\mathcal{D}) = D(\mathcal{B}) \subseteq l^p(\mathbb{Z}; E) \to l^p(\mathbb{Z}; E),$$

$$(\mathcal{D}u)(k) = u(k) - Bu(k-1), \quad k \in \mathbb{Z}, \quad u(\cdot) \in D(\mathcal{B}).$$

Proposition 4.5 (see [49]). *Let the operator*

$$\mathcal{L} = -\frac{d}{dt} + A : D(\mathcal{L}) \subseteq \Upsilon(\mathbb{R}; E) \to \Upsilon(\mathbb{R}; E)$$

be invertible. Then the operator $\mathcal{D} : D(\mathcal{D}) \subseteq \Upsilon(\mathbb{Z}; E) \to \Upsilon(\mathbb{Z}; E)$ *is also invertible and*

$$\|\mathcal{D}^{-1}\| \leq 1 + C(\Upsilon)\big(M + M^2\|\mathcal{L}^{-1}\|\big).$$

Conversely, if $\mathcal{D}$ *is invertible, then* $\mathcal{L} : D(\mathcal{L}) \subseteq \Upsilon(\mathbb{R}; E) \to \Upsilon(\mathbb{R}; E)$ *is also invertible and*

$$\|\mathcal{L}^{-1}\| \leq C(\Upsilon)\big(M + M^2\|\mathcal{D}^{-1}\|\big).$$

Theorem 4.10 (see [45, 48]). *The difference operator*

$$(\mathcal{D}u)(k) = u(k) - Bu(k-1), \quad k \in \mathbb{N}, \quad u(\cdot) \in l^p(\mathbb{R}; E), \quad 1 \leq p \leq \infty,$$

is invertible if and only if

$$\sigma(B) \cap \mathbb{T} = \varnothing. \tag{4.6.7}$$

If the condition (4.6.7) is fulfilled, then the inverse operator has the form

$$(\mathcal{D}^{-1}v)(k) = \sum_{m \in \mathbb{Z}} \Gamma(k-m)v(m), \tag{4.6.8}$$

where $v(\cdot) \in l^p(\mathbb{Z}; E)$, $1 \leq p \leq \infty$, *and the function* $\Gamma(\cdot) : \mathbb{Z} \to B(E)$ *is defined by the formula*

$$\Gamma(k) = \begin{cases} B^k(I - P), & k \geq 0; \\ -B_0^{-k}P, & k \leq -1, \end{cases} \tag{4.6.9}$$

where B_0 *is the restriction of* B *to* $\mathcal{R}(P)$.

Definition 4.3. We say that an operator $B \in B(E)$ has an *exponential discrete dichotomy* with data (M, r, P) if $P \in B(E)$ is a projector in E and M and r are constants, $0 \leq r < 1$, such that the following properties hold:

(i) $B^k P = P B^k$ for all $k \in \mathbb{N}$;
(ii) $\big\|B^k(I - P)\big\| \leq Mr^k$ for all $k \in \mathbb{N}$;
(iii) $\hat{B} := B\big|_{\mathcal{R}(P)} : \mathcal{R}(P) \mapsto \mathcal{R}(P)$ is a homeomorphism satisfying the estimate

$$\|\hat{B}^{-k}P\| \leq Mr^k, \quad k \in \mathbb{N}.$$

The following result relates constants in the exponential discrete dichotomy for estimates of the resolvent $(\lambda I - B)^{-1}$ for $\lambda \in \mathbb{T}$.

Theorem 4.11 (see [223]). *For* $B \in B(E)$, *the following conditions are equivalent:*

(i) $\lambda \in \rho(B)$ *for all* $\lambda \in \mathbb{T}$ *and*

$$\left\| (\lambda I - B)^{-1} \right\| \leq \beta < \infty \quad \forall \lambda \in \mathbb{T}; \tag{4.6.10}$$

(ii) B *has an exponential dichotomy with data* (M, r, P).

More precisely, (i) *implies* (ii) *with*

$$M = \frac{2\beta^2}{\beta - 1}, \quad r = 1 - \frac{1}{2\beta}.$$

Conversely, (ii) *implies* (i) *with* $\beta = M \dfrac{1 + r}{1 - r}.$

Proof. Let (i) hold; without loss of generality, assume that $\beta > 1$. For $z \in \mathbb{C}, z \neq 0$, we have

$$\left| z - \frac{z}{|z|} \right| = |1 - |z||.$$

Hence, if $|1 - |z|| \beta < 1$, then the classical perturbation estimate shows that $z \in \rho(B)$ and

$$\left\| (zI - B)^{-1} \right\| \leq \frac{\beta}{1 - \beta|1 - |z||}. \tag{4.6.11}$$

We define P as the Riesz projector defined by the formula

$$I - P = \frac{1}{2\pi i} \int_{|z|=1} (zI - B)^{-1} dz. \tag{4.6.12}$$

Since B commutes with the resolvent, the condition (i) of Definition 4.3 holds. Further, using (4.6.11) and the Cauchy theorem, we can shift the contour:

$$I - P = \frac{1}{2\pi i} \int_{|z|=r} (zI - B)^{-1} dz \quad \text{for} \quad |1 - r| < \frac{1}{\beta}. \tag{4.6.13}$$

We claim that

$$B^k (I - P) = \frac{1}{2\pi i} \int_{|z|=r} z^k (zI - B)^{-1} dz \quad \text{for} \quad |1 - r| < \frac{1}{\beta}. \tag{4.6.14}$$

For $k = 0$, this follows from (4.6.13). If (4.6.14) holds for some k, we obtain

$$B^{k+1} P = \frac{1}{2\pi i} \int_{|z|=r} (B - zI + zI) z^k (zI - B)^{-1} dz$$

$$= \frac{1}{2\pi i} \int_{|z|=r} z^{k+1} (zI - B)^{-1} dz - \frac{1}{2\pi i} \int_{|z|=r} z^k dz$$

and thus the assertion holds for $k + 1$. Equations (4.6.14) and (4.6.11) immediately lead to the first dichotomy estimate for $1 - 1/\beta < r \leq 1$:

$$\|B^k P\| \leq \frac{1}{2\pi} 2\pi r r^k \frac{\beta}{1 - \beta(1 - r)} = \frac{\beta r^{k+1}}{1 - \beta(1 - r)} \quad \text{for } k \geq 0. \quad (4.6.15)$$

For the second dichotomy estimate, we use the resolvent equation

$$(zI - B)^{-1} = \frac{1}{z} I + z^{-1} B(zI - B)^{-1}. \quad (4.6.16)$$

For $|1 - r| < 1/\beta$, Eqs. (4.6.13) and (4.6.16) lead to the relation

$$I - P = \frac{1}{2\pi i} \int_{|z|=1} \left(\frac{1}{z} I - (zI - B)^{-1} \right) dz = -\frac{1}{2\pi i} \int_{|z|=r} \frac{1}{z} B(zI - B)^{-1} dz.$$

This shows that for $k = 1$, the following equality holds:

$$I - P = -B^k \frac{1}{2\pi i} \int_{|z|=r} z^{-k} (zI - B)^{-1} dz, \quad k \geq 1. \quad (4.6.17)$$

If (4.6.17) is known for some k, then we use (4.6.16) and find

$$I - P = -B^k \frac{1}{2\pi i} \int_{|z|=r} \left(z^{-(k+1)} + z^{-(k+1)} B(zI - B)^{-1} \right) dz$$

$$= -B^{k+1} \frac{1}{2\pi i} \int_{|z|=r} z^{-(k+1)} (zI - B)^{-1} dz.$$

We apply (4.6.17) to $x \in E$, use the fact that $I - P$ commutes with B and the estimate (4.6.11):

$$\|(I - P)x\| = \left\| \frac{1}{2\pi i} \int_{|z|=r} z^{-k} (zI - B)^{-1} dz B^k (I - P)x \right\|$$

$$\leq \frac{r^{-k}}{2\pi} 2\pi r \frac{\beta}{1 - \beta(r - 1)} \|B^k (I - P)x\|.$$

Finally, for all $k \geq 1$, $1 \leq r < 1 + 1/\beta$, and $u \in E$ we have

$$\|(I - P)u\| \leq \frac{\beta r^{-k+1}}{1 - \beta(r - 1)} \|B^k (I - P)u\|. \quad (4.6.18)$$

For $k = 1$, this estimate shows that the operator $\hat{B} = B_{|\mathcal{N}(P)} : \mathcal{N}(P) \mapsto \mathcal{N}(P)$ is one-to-one and has a bounded inverse operator. To show that $\hat{B}$ is surjective, we take $f \in \mathcal{N}(P)$ and set

$$v = -\frac{1}{2\pi i} \int_{|z|=1} z^{-1} (zI - B)^{-1} f \, dz.$$

From this equation we have $(I - P)v = 0$; using (4.6.16) we find

$$Bv = \frac{1}{2\pi i} \int_{|z|=1} \left(z^{-1}I - (zI - B)^{-1} \right) f \, dz = (I - P)f = f.$$

Therefore, $\hat{B}$ is a linear homeomorphism on $\mathcal{N}(P)$ satisfying the estimate

$$\left\| \hat{B}^{-k}(I - P)u \right\| \le \frac{\beta r^{-k+1}}{1 - \beta(r - 1)} \left\| (I - P)u \right\| \quad \text{for } 1 \le r < 1 + \frac{1}{\beta}. \quad (4.6.19)$$

This proves the exponential dichotomy.

To obtain specific values of constants, we choose $r = 1 - 1/(2\beta)$ in (4.6.15) and obtain the estimate $(2\beta - 1)r^k$. In order to have the same rate in the opposite direction, we apply (4.6.19) with

$$r = \left(1 - \frac{1}{2\beta} \right)^{-1} < 1 + \frac{1}{\beta}.$$

In (4.6.19), we then find the upper bound Mr^{-k}, where $M = (2\beta^2)/(\beta - 1)$. Since $M > 2\beta - 1$, we obtain the required result.

Now we assume the exponential dichotomy and prove the condition (i). For $|\lambda| = 1$, the equation $(\lambda I - B)u = f$ is equivalent to the system

$$\left(\lambda I - BP \right)Pu = Pf, \quad \left(\lambda I - \hat{B} \right)(I - P)u = (I - P)f,$$

which we can rewrite as follows:

$$\begin{aligned} \left(I - \lambda^{-1}BP \right)Pu &= \lambda^{-1}Pf, \\ \left(I - \lambda\hat{B}^{-1} \right)(I - P)u &= -\hat{B}^{-1}(I - P)f. \end{aligned} \quad (4.6.20)$$

Both equations have a unique solution given by a geometric series

$$\begin{aligned} Pu &= \sum_{k=0}^{\infty} \lambda^{-(k+1)} B^k Pf, \\ (I - P)u &= -\sum_{k=0}^{\infty} \hat{B}^{-(k+1)} \lambda^k (I - P)f. \end{aligned} \quad (4.6.21)$$

Then the exponential dichotomy implies the estimates

$$\|Pu\| \le \frac{M}{1 - r}\|f\|, \quad \|(I - P)u\| \le M\frac{r}{1 - r}\|f\|.$$

By the triangle inequality we obtain the condition (i) with $\beta = M\dfrac{1 + r}{1 - r}$. $\quad \square$

4.6.2 Discretization of dichotomy semigroups

Now we consider the case where the resolvent of the operator A is not necessarily compact.

Theorem 4.12 (see [66]). *Let the operators $\lambda I_n - B_n \in B(E_n)$ be Fredholm operators with $\mathrm{ind}(\lambda I_n - B_n) = 0$ for any $\lambda \in \mathbb{T}$, $n \in \mathbb{N}$. We also assume that $B \in B(E)$ possesses the property $\mathbb{T} \cap \sigma(B) = \varnothing$ and*

$$\lambda I_n - B_n \xrightarrow{\ \mathcal{PP}\ } \lambda I - B \quad \text{regularly for any } \lambda \in \mathbb{T}.$$

Then

$$\lambda I_n - B_n \xrightarrow{\ \mathcal{PP}\ } \lambda I - B \quad \text{stably for any } \lambda \in \mathbb{T}$$

and

$$\sup_{\lambda \in \mathbb{T}} \left\| \left(\lambda I_n - B_n \right)^{-1} \right\| < \infty.$$

Proof. Assume that there are some sequences $\{\lambda_n\}$, $\lambda_n \in \mathbb{T}$, and $\{x_n\}$, $x_n \in E_n$, such that $\|x_n\| = 1$ and $(\lambda_n I_n - B_n)x_n \xrightarrow{\ \mathcal{P}\ } 0$ as $n \to \infty$. Since $\mathbb{T}$ is compact, one can find $\mathbb{N}' \subset \mathbb{N}$ such that $\lambda_n \to \lambda_0 \in \mathbb{T}$ as $n \in \mathbb{N}'$. Moreover,

$$\lambda_0 I_n - B_n \xrightarrow{\ \mathcal{PP}\ } \lambda_0 I - B \quad \text{regularly}$$

for such $\lambda_0 \in \mathbb{T}$ and

$$\left(\lambda_0 I_n - B_n \right)x_n = \left(\lambda_0 I_n - \lambda_n I_n \right)x_n + \left(\lambda_n I_n - B_n \right)x_n \xrightarrow{\ \mathcal{P}\ } 0 \quad \text{for } n \in \mathbb{N}'.$$

Therefore, there is $\mathbb{N}'' \subset \mathbb{N}'$ such that $x_n \xrightarrow{\ \mathcal{P}\ } x_0 \neq 0$ as $n \in \mathbb{N}''$. But in this case

$$\left(\lambda_0 I_n - B_n \right)x_n \xrightarrow{\ \mathcal{P}\ } \left(\lambda_0 I - B \right)x_0 = 0 \quad \text{for } n \in \mathbb{N}'',$$

which contradicts the assumption $\mathbb{T} \cap \sigma(B) = \varnothing$. $\qquad\square$

Definition 4.4. We say that an operator $B_n \in B(E_n)$ *has the uniform exponential discrete dichotomy with data (M, r, P_n)* if $P_n \in B(E_n)$ *is a projector in E_n and M and r are constants, $0 \leq r < 1$, such that the following properties hold:*

(i) $B_n^k P_n = P_n B_n^k$ and $\|P_n\| \leq \text{const}$ for all $k, n \in \mathbb{N}$;

(ii) $\left\| B_n^k (I_n - P_n) \right\| \leq M r^k$ for all $k, n \in \mathbb{N}$;

(iii) $\hat{B}_n := B_n \big|_{\mathcal{R}(P_n)} : \mathcal{R}(P_n) \mapsto \mathcal{R}(P_n)$ is a homeomorphism satisfying the estimate

$$\left\| \hat{B}_n^{\,-k} P_n \right\| \leq M r^k, \quad k, n \in \mathbb{N}.$$

Theorem 4.13 (see [66]). *The following conditions are equivalent:*

(i) $\lambda I_n - B_n \xrightarrow{\mathcal{PP}} \lambda I - B$ *stably and* $\lambda \in \rho(B)$ *for any* $\lambda \in \mathbb{T}$;

(ii) *the operator* $\mathcal{D} = I - \mathcal{B}$ *is invertible and* $\mathcal{D}_n \xrightarrow{\mathcal{PP}} \mathcal{D}$ *stably, where* $(\mathcal{B}u)(k) = Bu(k-1)$, $k \in \mathbb{N}$;

(iii) $B_n \xrightarrow{\mathcal{PP}} B$, *the operator* $\lambda I - B$ *be invertible for any* $\lambda \in \mathbb{T}$, *and the operators* B_n *have the exponential discrete dichotomy with data* (M, r, P_n) *uniformly in* $n \in \mathbb{N}$.

Proof. The equivalence (i)$\Leftrightarrow$(iii) follows from Theorem 4.11. Indeed, by the formula (4.6.12) we see from (i) that $P_n \xrightarrow{\mathcal{PP}} P$ and $\|P_n\| \le$ const. By Theorem 4.12, we have

$$\sup_{\lambda \in \mathbb{T}} \left\| \left(\lambda I_n - B_n \right)^{-1} \right\| < \infty$$

and by Theorem 4.11(ii) we have (iii). Conversely, due to Theorem 4.11(i), it follows from the condition (iii) that $\lambda I_n - B_n \xrightarrow{\mathcal{PP}} \lambda I - B$ stably for all $\lambda \in \mathbb{T}$.

To prove the implication (ii)$\Rightarrow$(i), we note that (ii) means that for any $u(\cdot) \in l^p(\mathbb{Z}; E)$ we have

$$\sum_{k=-\infty}^{\infty} \left\| p_n u(k) - B_n p_n u(k-1) - p_n u(k) + p_n B u(k-1) \right\|_{E_n}^p \to 0 \quad \text{as } n \to \infty,$$

i.e., $B_n \xrightarrow{\mathcal{PP}} B$. Now we assume that the operator $I_n - B_n$ is not uniformly invertible in $l^\infty(\mathbb{Z}; E_n)$, i.e., for some sequence $\|x_n\| = 1$ one has

$$\left(\lambda_0 I_n - B_n \right) x_n \xrightarrow{\mathcal{P}} 0 \quad \text{as } n \to \infty \text{ for } \lambda_0 = 1 \in \mathbb{T}.$$

This means that for a stationary sequence $u_n(k) = x_n$, $k \in \mathbb{Z}$, $n \in \mathbb{T}$, we have

$$\left(\mathcal{D}_n u_n \right)(k) = u_n(k) - B_n u_n(k-1) = x_n - B_n x_n \xrightarrow{\mathcal{P}} 0 \quad \text{as } n \to \infty$$

for any $k \in \mathbb{N}$. But $\mathcal{D}_n \xrightarrow{\mathcal{PP}} \mathcal{D}$ stably, i.e.,

$$\left\| \mathcal{D}_n u_n \right\|_{l^\infty(\mathbb{Z}; E_n)} \ge \gamma \|u_n\|_{l^\infty(\mathbb{Z}; E_n)},$$

which contradicts the condition

$$\left(\lambda_0 I_n - B_n \right) x_n \xrightarrow{\mathcal{P}} 0 \quad \text{as } n \to \infty.$$

Now we show that $\mathcal{R}(\lambda_0 I_n - B_n) = E_n$. For any $y_n \in E_n$, $\|y_n\| = 1$, consider $v_n(k) = y_n$, $k \in \mathbb{Z}$, $n \in \mathbb{N}$. The solution of the equation $\mathcal{D}_n u_n = v_n$ is a sequence $u_n(k)$, which is also stationary, i.e., $(\lambda_0 I_n - B_n)x_n = y_n$, where $x_n = u_n(k)$, $k \in \mathbb{Z}$, $n \in \mathbb{N}$. To prove the implication (i)$\Rightarrow$(ii), we note that

$$\|\mathcal{D}_n\|_{B(l^p(\mathbb{Z};E_n))} \le \text{const}, \quad n \in \mathbb{N}.$$

Now for any $u(\cdot) \in l^p(\mathbb{Z}; E)$ and any $\epsilon > 0$, one can find $K \in \mathbb{N}$ such that

$$\left(\sum_{k=K}^{\infty} + \sum_{k=-K}^{-\infty} \right) \|u(k)\|^p \le \epsilon.$$

Moreover,

$$\sum_{k=-K}^{K} \left\| p_n u(k) - B_n p_n u(k-1) - p_n u(k) + p_n B u(k-1) \right\|_{E_n}^p \to 0 \quad \text{as } n \to \infty$$

since $B_n \xrightarrow{\mathcal{PP}} B$. Thus, we have $\mathcal{D}_n \xrightarrow{\mathcal{PP}} \mathcal{D}$. The convergence $\mathcal{D}_n^{-1} \xrightarrow{\mathcal{PP}} \mathcal{D}^{-1}$ follows from the formula (4.6.8). The theorem is proved. $\square$

4.7 General Dichotomy for Semidiscretization

In Sec. 3.1.1, the existence of an isolated hyperbolic equilibrium point $u^* = -A^{-1}f(u^*)$ followed from the compactness of the operator $A^{-1}f(\cdot)$ and the smoothness of the function $f(\cdot)$. In the case of a condensing operator $A^{-1}f(\cdot)$ with constant $q < 1$, the existence of an isolated hyperbolic equilibrium $u^* = -A^{-1}f(u^*)$ follows from the results of [3, Sec. 3.3.5] and the differentiability of $f(\cdot)$. As a consequence, the condensing property of $A^{-1}f(\cdot)$ can be derived, for example, as in [16, (9.3)]. If the operators $A_n^{-1}f_n(\cdot)$ are also condensing with constant $q < 1$, then the boundedness of $\|x_n\|$ and the $\mathcal{P}$-compactness of $\{x_n I_n - \lambda A_n^{-1}f_n(x_n)\}$ for any $0 < \lambda \le 1$ imply that $\{x_n\}$ is $\mathcal{P}$-compact. Hence, all conditions of [288, Theorem 4] are satisfied and, therefore, equilibrium points $\{u_n^*\}$ for (4.2.9) exist and possess the property $u_n^* \xrightarrow{\mathcal{P}} u^*$.

Now we are in a position to state our main result on uniform estimates for terms of the discrete solutions (4.2.16).

Theorem 4.14 (see [66]). *Let the operators A_n and A be generators of analytic C_0-semigroups and let Condition (B$_1$) be fulfilled. Assume also that the semigroup $e^{tA_{u^*}}$ is hyperbolic,*

$$\sigma(A_{u^*}) \cap \{\lambda : \operatorname{Re}\lambda \ge 0\} \subseteq P\sigma(A_{u^*}), \quad \dim P(\sigma+) < \infty,$$

and for $\rho > 0$ such that

$$\{\lambda : -\rho \le \operatorname{Re}\lambda \le \rho\} \subset \rho(A_{u^*}),$$

the operators $\lambda I_n - A_{u_n^*,n}$ are Fredholm operators of index 0 and the opera-
tors $\lambda I_n - A_{u_n^*,n}$ and $\lambda I - A_{u^*}$ are regularly consistent for any $\operatorname{Re} \lambda \geq -\rho$.
Then $P_n(\sigma_n^+) \xrightarrow{\mathcal{PP}} P(\sigma^+)$ compactly and

$$
\begin{cases}
\left\| e^{tA_{u_n^*,n}} \left(I_n - P_n(\sigma_n^+) \right) \right\|_{E_n} \leq M_2 e^{-\gamma t}, & t \geq 0, \\[2mm]
\left\| e^{tA_{u_n^*,n}} P_n(\sigma_n^+) \right\|_{E_n} \leq M_2 e^{\gamma t}, & t \leq 0,
\end{cases}
\tag{4.7.1}
$$

where $\gamma > 0$.

Proof. Condition (B_1) implies that

$$
(\lambda I_n - A_{u_n^*,n})^{-1} \xrightarrow{\mathcal{PP}} (\lambda I - A_{u^*})^{-1}
\tag{4.7.2}
$$

for $-\rho \leq \operatorname{Re} \lambda \leq \rho$ and sufficiently large $|\lambda|$. For other λ satisfying the condition $-\rho \leq \operatorname{Re} \lambda \leq \rho$, the convergence (4.7.2) follows from a theorem similar to Theorem 3.4 for closed operators. Now the compact convergence $P_n(\sigma_n^+) \xrightarrow{\mathcal{PP}} P(\sigma^+)$ can be obtained in the same way as in [235] and the estimates (4.7.1) are obtained as in [56, 69]. Theorem 4.7.1 is proved. □

Remark 4.5. Certainly, Theorem 4.14 also holds for the case of any oper-ator A, which generates an analytic hyperbolic C_0-semigroup and satisfies the conditions $\sigma(A) \cap \{\lambda : \operatorname{Re} \lambda \geq 0\} \subseteq P\sigma(A)$ and $\dim P(\sigma+) < \infty$ and the corresponding conditions on the approximations of the operators. The structure of operators like $A + f_u'(u^*)$ is irrelevant.

Theorem 4.14 implies the following assertion.

Corollary 4.2. *Let the operators A_n and A be the generators of analytic C_0-semigroups and let Condition (B_1) be fulfilled. Assume also that the C_0-semigroup $e^{tA_{u^*}}$ is hyperbolic,*

$$
\sigma(A_{u^*}) \cap \{\lambda : \operatorname{Re} \lambda \geq 0\} = P\sigma(A_{u^*}), \quad \dim P(\sigma+) < \infty.
$$

Moreover, let $\Delta_{cc}(A_n, A) \neq \varnothing$ and the resolvents of the operators A_n and A be compact operators. Then the estimates (4.7.1) hold.

Proof. We set

$$
B_n = e^{1A_{u_n^*,n}}, \quad B = e^{1A_{u^*}}.
$$

It is known (see [235]) that the condition $\Delta_{cc} \neq \varnothing$ is equivalent to the compact convergence $B_n \xrightarrow{\mathcal{PP}} B$. Then the condition (i) of Theorem 4.13 is fulfilled and one gets the discrete dichotomy for B_n. On the other hand, since $B_n = e^{1A_{u_n^*,n}}$, this means, by Theorem 4.11 and Proposition 4.5, that the operators $\mathcal{L}_n = -\dfrac{d}{dt} + A_{u_n^*,n}$ are invertible and thus, by Theorem 4.9, the estimates (4.7.1) are valid.   □

4.7.1   Dichotomy for condensing operators in semidiscretization

Here we use the same notation as in Theorem 3.5.

Definition 4.5. We say that operators $B_n \in B(E_n)$ are *jointly condensing with constant $q > 0$ with respect to a measure* $\mu(\cdot)$ if for any bounded sequence $\{x_n\}$, $x_n \in E_n$, one has

$$\mu(\{B_n x_n\}) \leq q\mu(\{x_n\}).$$

It is known (see [3, p. 82]) that outside a closed disk of radius q centered at zero, each operator B_n has only isolated points of spectrum, each of which can be only an eigenvalue of finite multiplicity.

Proposition 4.6 (see [223]). *Let $B_n \xrightarrow{\mathcal{PP}} B$ for $B_n \in B(E_n)$ and $B \in B(E)$ and let*

$$\mu(\{B_n x_n\}) \leq q\mu(\{x_n\})$$

for any bounded sequence $\{x_n\}$, $x_n \in E_n$. Assume that $\sigma(B) \cap \Psi = \varnothing$, where $\Psi \subset \mathbb{C} \setminus \{\lambda : |\lambda| \leq q\}$ is a bounded closed set and $\sigma(B) \setminus \{\lambda : |\lambda| \leq q\}$ consists only of points of the discrete spectrum. Then there is a constant $C > 0$ such that $\left\| (\lambda I_n - B_n)^{-1} \right\| \leq C$, $\lambda \in \Psi$, $n \in \mathbb{N}$.

Proof. Any point $\lambda \in \Psi$ belongs to $P\sigma(B_n) \cup \rho(B_n)$. This means that for a sequence $\|x_n\| = 1$, $x_n \in E_n$, the following two cases are possible:

$$(\lambda I_n - B_n)x_n = 0 \quad \text{or} \quad \left\| (\lambda I_n - B_n)x_n \right\| \geq \gamma_{\lambda,n}\|x_n\|$$

with some $\gamma_{\lambda,n} > 0$ and $\lambda \in \Psi$. We show that in fact we have

$$\left\| (\lambda I_n - B_n)x_n \right\| \geq \gamma_\Psi \|x_n\|, \quad \lambda \in \Psi.$$

On the contrary, assume that there are sequences $\{\lambda_n\}$, $\lambda_n \in \Psi$, and $\{x_n\}$, $\|x_n\| = 1$, such that

$$(\lambda_n I_n - B_n)x_n \xrightarrow{\mathcal{P}} 0 \quad \text{for } n \in \mathbb{N}.$$

Then $\lambda_n \to \lambda_0 \in \Psi$, $n \in \mathbb{N}' \subseteq \mathbb{N}$. For $\tilde{r} = \inf\{|\xi| : \xi \in \Psi\}$ we have

$$\mu(\{x_n\}) \leq \frac{|\lambda_n|}{\tilde{r}}\mu(\{x_n\}) \leq \frac{\mu(\{B_n x_n\})}{\tilde{r}} \leq \frac{q}{\tilde{r}}\mu(\{x_n\}),$$

which, due to the inequality $q/\tilde{r} < 1$, implies that $\mu(\{x_n\}) = 0$, i.e., $\{x_n\}$ is $\mathcal{P}$-compact. Now

$$x_n \xrightarrow{\mathcal{P}} x_0, \quad n \in \mathbb{N}'' \subseteq \mathbb{N}',$$

$$B_n x_n \xrightarrow{\mathcal{P}} B x_0, \quad \lambda_n x_n \xrightarrow{\mathcal{P}} \lambda_0 x_0, \quad n \in \mathbb{N}'',$$

i.e., $\lambda_0 x_0 = B x_0$ with $\|x_0\| = 1$, which contradicts the assumption $\sigma(B) \cap \Psi = \varnothing$. Proposition 4.6 is proved. $\qquad\square$

Let us denote by $\nu(\cdot)$ the noncompactness measure in E.

Proposition 4.7 (see [223]). *Let* $B_n \xrightarrow{\mathcal{PP}} B$,

$$\mu(\{B_n x_n\}) \le q\mu(\{x_n\}) \quad \text{for any bounded sequence } \{x_n\},\ x_n \in E_n,$$
$$\nu(\{B x^n\}) \le q\nu(\{x^n\}) \quad \text{for any bounded sequence } \{x^n\},\ x^n \in E,$$

and $q < 1$. *Then any* $\lambda_0 \in \sigma(B)$, $|\lambda_0| > q$, *is an isolated eigenvalue with the finite-dimensional projector* $P(\lambda_0)$ *and there are a sequence* $\{\lambda_n\}$, $\lambda_n \in \sigma(B_n)$, *and a sequence of projectors* $P_n(\lambda_0) \in B(E_n)$ *such that* $\lambda_n \to \lambda_0$ *and* $P_n(\lambda_0) \xrightarrow{\mathcal{PP}} P(\lambda_0)$ *compactly.*

Proof. First, note that for $\Gamma_r = \{\lambda : |\lambda - \lambda_0| = r\} \subset \mathbb{C} \setminus \{\lambda : |\lambda| \le q\}$, where r can be taken sufficiently small, we have by Proposition 4.6

$$\left(\lambda I_n - B_n\right)^{-1} \xrightarrow{\mathcal{PP}} (\lambda I - B)^{-1} \quad \text{for } \lambda \in \Gamma_r,\ n \in \mathbb{N}.$$

Therefore, $P_n(\lambda_0) \xrightarrow{\mathcal{PP}} P(\lambda_0)$. To prove the compact convergence of these projectors, note that

$$\mu\left(\left\{\left(\lambda_0 I_n - B_n\right)x_n\right\}\right) \ge |\lambda_0|\mu(\{x_n\}) - \mu(\{B_n x_n\})$$
$$\ge |\lambda_0|\,\mu(\{x_n\}) - q\mu(\{x_n\}) \ge \gamma\mu(\{x_n\}),$$

where $\gamma = |\lambda_0| - q > 0$. This means that

$$\mu\left(B\left\{\left(\lambda_0 I_n - B_n\right)^k x_n\right\}\right) \ge \gamma^k \mu(\{x_n\}) \quad \text{for any } k \in \mathbb{N}. \tag{4.7.3}$$

Moreover,

$$\left(\lambda_0 I_n - B_n\right)^k P_n(\lambda_0)x_n = \frac{1}{2\pi i} \int_{\Gamma_r} \left(\lambda_0 - \lambda\right)^k \left(\lambda I_n - B_n\right)^{-1} x_n d\lambda.$$

From this representation and (4.7.3) we have

$$\gamma^k \mu\left(\{P_n(\lambda_0)x_n\}\right) \le \left\|\left(\lambda_0 I_n - B_n\right)^k P_n(\lambda_0)x_n\right\| \le \frac{C}{2\pi} r^k \|x_n\|.$$

Clearly, it follows from $r/\gamma < 1$ that $(r/\gamma)^k \to 0$ as $k \to \infty$. This means that $P_n(\lambda_0) \xrightarrow{\mathcal{PP}} P(\lambda_0)$ compactly. Proposition 4.7 is proved. $\qquad\square$

Theorem 4.15 (see [223]). *Let Conditions* (A) *and* (B$_1$) *be fulfilled, the analytic* C_0-*semigroup* e^{tA}, $t \in \overline{\mathbb{R}}^+$, *be hyperbolic such that the set* $\sigma(A) \cap \{\lambda : \operatorname{Re}\lambda \ge 0\}$ *consists of a finite number of points* $P\sigma(A)$, *and let* $\dim P(\sigma+) < \infty$. *Assume also that* $\mu(\{B_n x_n\}) \le q\mu(\{x_n\})$ *for any bounded sequence* $\{x_n\}$, $x_n \in E_n$, *with* $q < 1$, *where* $B_n = e^{1 A_n}$. *Then the estimates* (4.7.1) *hold and* $P_n(\sigma_n+) \xrightarrow{\mathcal{PP}} P(\sigma+)$ *compactly.*

Proof. Owing to the spectral mapping theorem, the spectrum of the operator $B = e^{1A}$, which is located outside of the unit disc $\mathbb{T}$, consists of finite number of points of the set

$$P\sigma(e^{1A}) = \Big\{ \zeta : \ \zeta = e^{\lambda}, \ \lambda \in P\sigma(A) \cap \{\xi : \operatorname{Re}\xi \geq 0\} \Big\}.$$

Moreover, since $q < 1$, for any set $\Psi \supset \mathbb{T}$ and $\Psi \subset \rho(B)$, $B = e^{1A}$, one has $\big\|(\lambda I_n - B_n)^{-1}\big\| \leq$ const for $\lambda \in \Psi$ by Proposition 4.7. Now Theorem 4.13 implies that the operators B_n have the discrete dichotomy. By Theorem 4.10 the operators $\mathcal{D}_n$ are invertible and by Proposition 4.5 and Theorem 4.9 we get that the semigroups e^{tA_n}, $t \in \mathbb{R}_+$, have the exponential dichotomy uniformly in the index $n \in \mathbb{N}$ and hence the estimates (4.7.1) hold. By Proposition 4.7, $P_n(\lambda_0) \xrightarrow{\mathcal{PP}} P(\lambda_0)$ compactly, which implies that the conditions of Theorem 4.14 are satisfied. Theorem 4.15 is proved.

$\square$

4.8 General Dichotomy for Discretization in Time Variable

Now we consider the discretization of the problem (4.2.11) in time by the following scheme:

$$\frac{V_n(t + \tau_n) - V_n(t)}{\tau_n} = A_{u_n^*,n} V_n(t + \tau_n) + F_{u_n^*,n}(V_n(t)), \quad t = k\tau_n, \quad (4.8.1)$$

with the initial data $V_n(0) = v_n^0$. The solution of such problem is given by the formula

$$V_n(t + \tau_n) = \big(I_n - \tau_n A_{u_n^*,n}\big)^{-1} V_n(t) + \tau_n\big(I_n - \tau_n A_{u_n^*,n}\big)^{-1} F_{u_n^*,n}(V_n(t))$$

$$= \big(I_n - \tau_n A_{u_n^*,n}\big)^{-k} V_n(0)$$

$$+ \tau_n \sum_{j=0}^{k} \big(I_n - \tau_n A_{u_n^*,n}\big)^{-(k-j+1)} F_{u_n^*,n}\big(V_n(j\tau_n)\big), \quad t = k\tau_n,$$

where $V_n(0) = v_n^0$.

The problem (4.2.15) also can be discretized by the scheme (4.8.1), so we have (4.4.6). The solution of the problem (4.4.6) can be obtained by the formulas

$$(I_n - P_n)V_n(t + \tau_n) = \big(I_n - \tau_n A_{u_n^*,n}\big)^{-1}(I_n - P_n)V_n(t)$$

$$+ \tau_n\big(I_n - \tau_n A_{u_n^*,n}\big)^{-1}(I_n - P_n)F_{u_n^*,n}\big(V_n(j\tau_n)\big),$$

and

$$\big(I_n - \tau_n A_{u_n^*,n}\big)P_n V_n(t + \tau_n) = P_n V_n(t) + \tau_n P_n F_{u_n^*,n}\big(V_n(k\tau_n)\big), \quad t = k\tau_n.$$

We can represent a solution of the problem (4.4.6) in the form (4.4). From (4.4) it is clear that the corresponding estimates of powers of the operators

$$\left(I_n - \tau_n A_{u_n^*,n}\right)^{-k+1}(I_n - P_n), \quad \left(I_n - \tau_n A_{u_n^*,n}\right)^{K-k} P_n$$

play a fundamental role in approximation of solutions of (4.1.6) in a neighborhood of u^*.

Theorem 4.16 (see [66]). *Let the operators A_n and A be generators of analytic C_0-semigroups and let Condition (B_1) be fulfilled. Assume also that the analytic C_0-semigroup $e^{tA_{u^*}}$, $t \in \overline{\mathbb{R}}^+$, is hyperbolic and for $\rho > 0$ such that $\{\lambda : -\rho \leq \mathrm{Re}\,\lambda \leq \rho\} \subset \rho(A)$, the operators $\lambda I_n - A_{u_n^*,n}$ are Fredholm operators of index 0 and the operators $\lambda I_n - A_{u_n^*,n}$ and $\lambda I - A_{u^*}$ are regularly consistent for any $\mathrm{Re}\,\lambda \geq -\rho$. Then $P_n(\sigma_n+) \xrightarrow{\ \mathcal{PP}\ } P(\sigma+)$ compactly and*

$$\begin{cases} \left\| \left(I_n - \tau_n A_{u_n^*,n}\right)^{-k_n}(I_n - P_n) \right\|_{E_n} \leq M_2 e^{-\gamma t}, & t \geq 0, \\[2mm] \left\| \left(I_n - \tau_n A_{u_n^*,n}\right)^{k_n} P_n \right\|_{E_n} \leq M_2 e^{\gamma t}, & t \leq 0, \end{cases} \qquad (4.8.2)$$

for some $\gamma > 0$.

Proof. Theorem 4.14 implies that the C_0-semigroups $e^{tA_{u_n^*,n}}$ have the dichotomy uniformly in $n \in \mathbb{N}$. As in [231], we can see that

$$\left\| \left(I_n - \tau_n A_{u_n^*,n}\right)^{-k_n} - e^{tA_{u_n^*,n}} \right\| \leq \frac{M}{t}\tau_n e^{\omega t},$$

where $t = k_n \tau_n = 1$. Using the perturbation dichotomy theorem from [129, p. 254], we conclude that (4.8.2) holds. The theorem is proved. $\quad\square$

4.8.1   Discretization in the time variable and the condensing property

Theorem 4.17 (see [66]). *Let Conditions (A) and (B_1) be fulfilled. Assume that $\mu(\{B_n x_n\}) \leq q\mu(\{x_n\})$ for any bounded sequence $\{x_n\}$, $x_n \in E_n$, where $q < 1$, and $B_n = e^{A_{u_n^*,n}}$. Assume also that the analytic C_0-semigroup $e^{tA_{u^*}}$, $t \in \overline{\mathbb{R}}^+$, is hyperbolic. Then*

$$\begin{cases} \left\| \left(I_n - \tau_n A_{u_n^*,n}\right)^{-k_n}(I_n - P_n) \right\|_{E_n} \leq M_2 r^{[t]}, & t = k_n \tau_n \geq 0, \\[2mm] \left\| \left(I_n - \tau_n A_{u_n^*,n}\right)^{k_n} P_n \right\|_{E_n} \leq M_2 r^{-[t]}, & t = -k_n \tau_n \leq 0, \end{cases} \qquad (4.8.3)$$

where $r < 1$.

Proof. Theorem 4.15 implies that (4.2.14) holds. By the perturbation theorem (see [129]), similarly to the proof of Theorem 4.16, we obtain (4.8.3). Theorem 4.17 is proved. $\qquad\square$

Remark 4.6. We have

$$\left\|F'_{u_n^*,n}(w_n)\right\| \le c_\rho \|w_n\|_{E_n^\theta},$$

where $c_\rho \to 0$ as $\rho \to 0$. Here $\rho > 0$ is the radius of balls $\mathcal{U}_{E_n^\theta}(0;\rho)$ for which there exists $\delta > 0$ such that

$$\sup_{n\in\mathbb{N}} \sup_{\|w_n\|_{E_n^\theta} \le \delta} \left\|f'_n\left(w_n + p_n^\theta u^*\right) - f'_n\left(p_n^\theta u^*\right)\right\|_{B(E_n^\theta,E_n)} \le \rho.$$

4.9 Semidiscrete Dichotomy for Condensing Operators

Proposition 4.8 (see [223]). *Assume that the conditions of Proposition 4.7 are satisfied with $B = e^{A_{u^*}}$ and $B_n = e^{A_{u_n^*,n}}$, where $q < 1$, and the semigroup $e^{tA_{u^*}}$, $t \in \mathbb{R}^+$, is hyperbolic. Then the conclusions of Theorem 4.14, in particular, (4.7.1), hold.*

4.9.1 *Shadowing for discretization in space and the condensing property*

If the resolvent of A is not compact, then we still get the shadowing property, but just on compact subsets of initial data $\mathcal{K} \subset \mathcal{U}_{E^\theta}(u^*;\rho)$ as formulated in the following theorem.

Theorem 4.18 (see [66]). *Let the operator A be the generator of an exponentially decreasing, analytic C_0-semigroup and u^* be a hyperbolic equilibrium point of the problem (4.1.5). Let*

$$\left\{\lambda:\ -\rho_1 \le \operatorname{Re}\lambda \le \rho_1\right\} \subset \rho(A),$$

the operators $\lambda I_n - A_{u_n^*,n}$ and $\lambda I - A_{u^*}$ be regularly consistent for any $\operatorname{Re}\lambda \ge -\rho_1$, and

$$f_n(x_n) \xrightarrow{\mathcal{P}} f(x), \quad f'_n(x_n) \xrightarrow{\mathcal{P}^\theta\mathcal{P}} f'(x) \quad as\ x_n \xrightarrow{\mathcal{P}^\theta} x.$$

Also assume that Conditions (B_1) and $(\mathrm{F1})$–$(\mathrm{F3})$ are fulfilled. Then there exists $\rho_0 > 0$ possessing the following property: for any compact set $\mathcal{K} \subset \mathcal{U}_{E^\theta}(u^*;\rho)$ and any $\epsilon_0 > 0$, there is a number $n(\epsilon_0) \in \mathbb{N}$ such that for any mild solution $u(t)$ of the problem (4.1.5) satisfying $u(\cdot) \in \mathcal{U}_{E^\theta}(u^*;\rho_0)$,

$0 \leq t \leq T$, *for some* $0 < T \leq \infty$ *and* $u(0) \in \mathcal{K}$ *there exist initial values* $u_n^0 \in E_n^\theta$, $n \geq n(\epsilon_0)$, *such that the mild solution* $u_n(t; u_n^0)$ *of the problem* (4.2.9) *exists on* $[0, T]$ *and satisfies the estimate*

$$\sup_{0 \leq t \leq T} \left\| u_n(t; u_n^0) - \tilde{p}_n^\theta u(t) \right\|_{E_n^\theta} \leq \epsilon_0 \quad \forall n \geq n(\epsilon_0). \tag{4.9.1}$$

Proof. Instead of the condition $\Delta_{cc} \neq \varnothing$ in Theorem 4.4, we use Condition (B_1) and the regular consistence of $\lambda I_n - A_n$ and $\lambda I - A$ in $\operatorname{Re}\lambda \geq -\rho_1$ to obtain the compact convergence $P_n \xrightarrow{\mathcal{PP}} P$ and the estimates (4.7.1). One can see that all statements in the proof of Theorem 4.4 hold. Since the set $\mathcal{K} \subset \mathcal{U}_{E^\theta}(0; \rho)$ is compact, the set of all orbits $\{u(t; u^0) : 0 \leq t \leq T \leq \infty, u^0 \in \mathcal{K}\} \cup \{u^*\}$ is compact in E. Then the proof is completed in the same way as for Theorem 4.4. $\qquad\square$

Theorem 4.19 (see [66]). *Let the operator A be the generator of an exponentially decreasing, analytic C_0-semigroup and u^* be a hyperbolic equilibrium point of the problem* (4.1.5). *Let*

$$\{\lambda : -\rho_1 \leq \operatorname{Re}\lambda \leq \rho_1\} \subset \rho(A),$$

the operators $\lambda I_n - A_{u_n^*, n}$ and $\lambda I - A_{u^*}$ be regularly consistent for any $\operatorname{Re}\lambda \geq -\rho_1$,

$$f_n(x_n) \xrightarrow{\mathcal{P}} f(x), \quad f_n'(x_n) \xrightarrow{\mathcal{P}^\theta \mathcal{P}} f'(x) \quad as \ x_n \xrightarrow{\mathcal{P}^\theta} x.$$

Also assume that and Conditions (B_1) and $(F1)$–$(F3)$ are fulfilled. Then there exists $\rho_0 > 0$ with the following property: for any compact set $\mathcal{K} \subset \mathcal{U}_{E^\theta}(u^*; \rho_0)$ and any $\epsilon_0 > 0$, there is a number $n(\epsilon_0) \in \mathbb{N}$ such that for any sequence $\{u_n(\cdot)\}$ of mild solutions of the problems (4.2.9) *with sequences $\{u_n^0\}$, $u_n(0) = u_n^0 \in \mathcal{K}_n = p_n^\theta \mathcal{K} \cap \mathcal{U}_{E_n^\theta}(u_n^*; \rho_0)$, there exist initial values $u^{n,0} \in \mathcal{K}$, $n \geq n(\epsilon_0)$, such that the mild solutions $u(t; u^{n,0})$ of the problems* (4.1.5) *exist on $[0, T]$ and satisfy the estimate*

$$\sup_{0 \leq t \leq T} \left\| u_n(t) - \tilde{p}_n^\theta u(t; u^{n,0}) \right\|_{E_n^\theta} \leq \epsilon_0 \quad \forall n \geq n(\epsilon_0). \tag{4.9.2}$$

Proof. According to [56, Lemma 4.5], there is a sequence $\{u^{n,0}\}$, $u^{n,0} \in \mathcal{K}$, such that

$$\left\| u_n^0 - \tilde{p}_n^\theta u^{n,0} \right\| \leq \epsilon_0 \quad \text{for } n \geq n(\epsilon_0).$$

Following the proof of Theorem 4.5, one can show that all conclusions of Theorem 4.4 are satisfied and (4.9.2) holds. $\qquad\square$

Theorem 4.20 (see [66]). *Let the operator A be the generator of an exponentially decreasing, analytic C_0-semigroup and u^* be a hyperbolic equilibrium point of the problem* (4.1.5). *Assume also that $\mu(\{B_n x_n\}) \leq q\mu(\{x_n\})$ for any bounded sequence $\{x_n\}$, $x_n \in E_n$, where $q < 1$, $B_n = e^{1A_n}$,*

$$f_n(x_n) \xrightarrow{\mathcal{P}} f(x), \quad f'_n(x_n) \xrightarrow{\mathcal{P}^\theta \mathcal{P}} f'(x) \quad \text{as } x_n \xrightarrow{\mathcal{P}^\theta} x,$$

and Conditions (B) *and* (F1)–(F3) *are satisfied. Then there exists $\rho_0 > 0$ with the following property: for any compact set $\mathcal{K} \subset \mathcal{U}_{E^\theta}(u^*; \rho_0)$ and any $\epsilon_0 > 0$, there is a number $n(\epsilon_0) \in \mathbb{N}$ such that for any mild solution $u(t)$ of the problem* (4.1.5) *satisfying the condition $u(\cdot) \in \mathcal{U}_{E^\theta}(u^*; \rho_0)$, $0 \leq t \leq T$, for some $0 < T \leq \infty$ and $u(0) \in \mathcal{K}$, there exist initial values $u_n^0 \in E_n^\theta$, $n \geq n(\epsilon_0)$, such that the mild solution $u_n(t; u_n^0)$ of the problem* (4.2.9) *exists on $[0, T]$ and satisfies the estimate* (4.9.1).

Proof. According to Theorem 4.15, the condition $\mu(\{B_n x_n\}) \leq q\mu(\{x_n\})$ implies the compact convergence $P_n \xrightarrow{\mathcal{PP}} P$. Hence due to the estimate (4.7.1), it suffices to repeat the proof of Theorem 4.18. □

Theorem 4.21 (see [66]). *Let the operator A be the generator of an exponentially decreasing, analytic C_0-semigroup and u^* be a hyperbolic equilibrium point of the problem* (4.1.5). *Assume that $\mu(\{B_n x_n\}) \leq q\mu(\{x_n\})$ for any bounded sequence $\{x_n\}$, $x_n \in E_n$, where $q < 1$ and $B_n = e^{1A_n}$,*

$$f_n(x_n) \xrightarrow{\mathcal{P}} f(x), \quad f'_n(x_n) \xrightarrow{\mathcal{P}^\theta \mathcal{P}} f'(x) \quad \text{as } x_n \xrightarrow{\mathcal{P}^\theta} x.$$

Also let Conditions (B_1) *and* (F1)–(F3) *be fulfilled. Then there exists $\rho_0 > 0$ with the following property: for any compact set $\mathcal{K} \subset \mathcal{U}_{E^\theta}(u^*; \rho_0)$ and any $\epsilon_0 > 0$, there exists $n_0(\epsilon_0) \in \mathbb{N}$ such that for any sequence $\{u_n(\cdot)\}$ of mild solutions of the problems* (4.2.9) *with the sequences $\{u_n^0\}$, $u_n(0) = u_n^0 \in \mathcal{K}_n = p_n^\theta \mathcal{K} \cap \mathcal{U}_{E_n^\theta}(u_n^*; \rho_0)$, there exist initial values $u^{n,0} \in \mathcal{K}$, $n \geq n_0$, such that the mild solutions $u(t; u^{n,0})$ of the problem* (4.1.5) *exist on $[0, T]$ and satisfy the estimate* (4.9.2).

Proof. We complete the proof by using the fact that the condensing estimates $\mu(\{B_n x_n\}) \leq q\mu(\{x_n\})$ with $q < 1$, which are uniform in $n \in \mathbb{N}$, implies the dichotomy as in (4.7.1). Our statements follow as in Theorem 4.19. □

4.9.2 *Shadowing for discretizing in time variable and condensing property*

Lemmas 4.4–4.6 are valid under the conditions of Theorems 4.16–4.17 since we have the discrete dichotomy and the compact convergence $P_n \xrightarrow{\mathcal{PP}} P$.

Theorem 4.22 (see [66]). *Let the operator A be the generator of an exponentially decreasing, analytic C_0-semigroup and u^* be a hyperbolic equilibrium point of the problem (4.1.5). Assume also that*

$$\{\lambda:\ -\rho_1 \le \operatorname{Re}\lambda \le \rho_1\} \subset \rho(A),$$

the operators $\lambda I_n - A_{u_n^*,n}$ and $\lambda I - A_{u^*}$ are regularly consistent for any $\operatorname{Re}\lambda \ge -\rho_1$,

$$f_n(x_n) \xrightarrow{\ \mathcal{P}\ } f(x), \quad f_n'(x_n) \xrightarrow{\ \mathcal{P}^\theta \mathcal{P}\ } f'(x) \quad as\ x_n \xrightarrow{\ \mathcal{P}^\theta\ } x,$$

and Conditions (B$_1$) and (F1)–(F3) are fulfilled. Then there exists $\rho_0 > 0$ with the following property: for any compact set $\mathcal{K} \subset \mathcal{U}_{E^\theta}(0;\rho)$ and any $\epsilon_0 > 0$, there is a number $n(\epsilon_0) \in \mathbb{N}$ such that for any mild solution $v(t)$ of the problem (4.1.9) satisfying the condition $v(t) \in \mathcal{U}_{E^\theta}(0;\rho_0)$, $0 \le t \le T$, for some $0 < T \le \infty$ and $v^0, v^T \in \mathcal{K} \cap \mathcal{U}_{E^\theta}(0;\rho)$, there exist elements $v_n^0, v_n^T \in E_n^\theta$, $n \ge n(\epsilon_0)$, such that the solutions $V_n(\cdot)$ of the problems (4.4.6) exist and satisfy the estimate

$$\sup_{0 \le t \le T} \left\| V_n\big(t; v_n^0, v_n^T\big) - \tilde{p}_n^\theta v\big(t; v^0, v^T\big) \right\|_{E_n^\theta} \le \epsilon_0 \quad for\ n \ge n(\epsilon_0). \qquad (4.9.3)$$

Proof. As in Theorem 4.7, we consider the solution (4.4) of (4.2.16). The dichotomy estimates follow from Theorem 4.16. $\qquad\square$

Theorem 4.23 (see [66]). *Let the operator A be the generator of exponentially decreasing, analytic C_0-semigroup and u^* be a hyperbolic equilibrium point of the problem (4.1.5). Assume also that*

$$\{\lambda:\ -\rho_1 \le \operatorname{Re}\lambda \le \rho_1\} \subset \rho(A),$$

the operators $\lambda I_n - A_{u_n^*,n}$ and $\lambda I - A_{u^*}$ are regularly consistent for any $\operatorname{Re}\lambda \ge -\rho_1$,

$$f_n(x_n) \xrightarrow{\ \mathcal{P}\ } f(x), \quad f_n'(x_n) \xrightarrow{\ \mathcal{P}^\theta \mathcal{P}\ } f'(x) \quad as\ x_n \xrightarrow{\ \mathcal{P}^\theta\ } x,$$

and Conditions (B$_1$) and (F1)–(F3) are fulfilled. Then there exists $\rho_0 > 0$ with the following property: for any compact set $\mathcal{K} \subset \mathcal{U}_{E^\theta}(0;\rho_0)$ and any $\epsilon_0 > 0$, there is a number $n(\epsilon_0) \in \mathbb{N}$ such that for any sequence $\{V_n(\cdot)\}$ of solutions of the problems (4.4.6) with sequences $\{v_n^0\}$ and $\{v_n^T\}$ from E_n^θ and $v_n^0, v_n^T \in \mathcal{K}_n = p_n^\theta \mathcal{K} \cap \mathcal{U}_{E_n^\theta}(0;\rho)$, there exist a set of elements $v^{n,0}, v^{n,T} \in E^\theta$ such that the mild solutions $v\big(t; v^{n,0}, v^{n,T}\big)$ of the problems (4.1.9) exist on $[0,T]$ and satisfy the estimates

$$\sup_{0 \le t \le T} \left\| V_n b\big(t; v_n^0, v_n^T\big) - \tilde{p}_n^\theta v\big(t; v^{n,0}, v^{n,T}\big) \right\|_{E_n^\theta} \le \epsilon_0 \quad for\ n \ge n(\epsilon_0). \qquad (4.9.4)$$

The proof of this theorem is based on Theorems 4.8 and 4.16.

Theorem 4.24 (see [66]). *Let the operator A be the generator of exponentially decreasing, analytic C_0-semigroup and u^* be a hyperbolic equilibrium point of the problem (4.1.5). Assume also that*

$$\mu(\{B_n x_n\}) \leq q\mu(\{x_n\})$$

for any bounded sequence $\{x_n\}$, $x_n \in E_n$, where $q < 1$, $B_n = e^{1A_n}$,

$$f_n(x_n) \xrightarrow{\mathcal{P}} f(x), \quad f_n'(x_n) \xrightarrow{\mathcal{P}^\theta \mathcal{P}} f'(x) \quad as \ x_n \xrightarrow{\mathcal{P}^\theta} x,$$

and Conditions (B_1) and $(F1)$–$(F3)$ are fulfilled. Then there exists $\rho_0 > 0$ with the following property: for any compact set $\mathcal{K} \subset \mathcal{U}_{E^\theta}(0; \rho)$ and any $\epsilon_0 > 0$, there is a number $n(\epsilon_0) \in \mathbb{N}$ such that for any mild solution $v(t)$ of the problem (4.1.9) satisfying the condition $v(t) \in \mathcal{U}_{E^\theta}(0; \rho_0)$, $0 \leq t \leq T$, for some $0 < T \leq \infty$ and $v^0, v^T \in \mathcal{K} \cap \mathcal{U}_{E^\theta}(0; \rho)$, there exist elements $v_n^0, v_n^T \in E_n^\theta$, $n \geq n(\epsilon_0)$, such that the solutions $V_n(\cdot)$ of the problems (4.4.6) exist and satisfies the estimate (4.9.3).

The proof of this theorem is based on Theorems 4.17 and 4.22.

Theorem 4.25 (see [66]). *Let the operator A be the generator of exponentially decreasing, analytic C_0-semigroup and u^* be a hyperbolic equilibrium point of the problem (4.1.5). Assume also that*

$$\mu(\{B_n x_n\}) \leq q\mu(\{x_n\})$$

for any bounded sequence $\{x_n\}$, $x_n \in E_n$, where $q < 1$, $B_n = e^{1A_n}$,

$$f_n(x_n) \xrightarrow{\mathcal{P}} f(x), \quad f_n'(x_n) \xrightarrow{\mathcal{P}^\theta \mathcal{P}} f'(x) \quad as \ x_n \xrightarrow{\mathcal{P}^\theta} x,$$

and Conditions (B_1) and $(F1)$–$(F3)$ are satisfied. Then there exists $\rho_0 > 0$ with the following property: for any compact set $\mathcal{K} \subset \mathcal{U}_{E^\theta}(0; \rho_0)$ and any $\epsilon_0 > 0$, there is a number $n(\epsilon_0) \in \mathbb{N}$ such that for any sequence $\{V_n(\cdot)\}$ of solutions of the problems (4.4.6) with sequences $\{v_n^0\}, \{v_n^T\}$ from E_n^θ and $v_n^0, v_n^T \in \mathcal{K}_n = p_n^\theta \mathcal{K} \cap \mathcal{U}_{E_n^\theta}(0; \rho)$, there exist a set of elements $v^{n,0}, v^{n,T} \in E^\theta$ such that the mild solutions $v(t; v^{n,0}, v^{n,T})$ of the problems (4.1.9) exist on $[0, T]$ and satisfy the estimates (4.9.4).

The proof of this theorem is based on Theorems 4.17 and 4.23.

Remark 4.7. In fact, we do not need to make the choice of elements v^T and v_n^T in connection to compact set $\mathcal{K}$ since the subspaces PE and $P_n E_n$ have finite dimensions and the operators p_n are uniformly invertible on $P_n E_n$ for $n \geq n_0$ with some n_0.

4.10   Verification of Condensing Conditions

The condition $\mu(B_n x_n) \leq q\mu(x_n)$, $q < 1$, in Theorems 4.15 and 4.17 can be checked, for example, in the case of the compact convergence of the operators $A_n^{-1} f_n'(u_n^*) \xrightarrow{\mathcal{PP}} A^{-1} f'(u^*)$. We present here an example where an analog of this condition is naturally satisfied.

Example 4.3. Consider in $L^2(\mathbb{R})$ the operator

$$(Av)(x) = v''(x) + av'(x) + bv(x), \quad x \in (-\infty, \infty).$$

Since $E = L^2(\mathbb{R})$, we can take

$$(p_n v)(x) = \frac{1}{h} \int_{-h/2}^{h/2} v(x + y)\, dy$$

and the main condition $\|p_n v\|_{L_h^2(Z)} \to \|v\|_{L^2(R)}$ is satisfied (see [288]).

As in [129, Sec. 5.4], we see that

$$\sigma_{\mathrm{ess}}(-A) \subset \left\{ \lambda : \ \mathrm{Re}\,\lambda - \frac{(\mathrm{Im}\,\lambda)^2}{a^2} \geq -b \right\}.$$

For the case of $a = 0$, we have $\sigma(A) \in (-\infty, b)$. So for $b < 0$ the operator A is a negative self-adjoint operator. Then we have the same for some difference scheme, say for central difference scheme:

$$A_n v_n(x) = \frac{v_n(x + h) - 2v_n(x) + v_n(x - h)}{h^2} + b\, v_n(x),$$

i.e., $\omega_{\mathrm{ess}}(A_n) \leq \omega_1 < 0$ uniformly in $h > 0$. Moreover, it is easy to see that

$$\|e^{tA_n}\| \leq M e^{\omega_2 t}, \quad t \geq 0, \quad \omega_2 < 0,$$

i.e.,

$$\mu\left(\left\{e^{tA_n} x_n\right\}\right) \leq \gamma \mu(\{x_n\}), \quad \gamma < 1, \quad \text{for some } t = t_0 > 0. \tag{4.10.1}$$

Now we introduce the notation $B = e^{t_0 A}$ and $B_n = e^{t_0 A_n}$. For analytic C_0-semigroups, the spectra of the operators A and B are strictly related, which is also concerned to the point spectra $P\sigma(B) = e^{t_0 P\sigma(A)}$. This means that for almost all n, the operators A_n have the spectra $\sigma(A_n) \cap \{\lambda : \ \mathrm{Re}\,\lambda > 0\}$ that approximate the spectrum $\sigma(A) \cap \{\lambda : \ \mathrm{Re}\,\lambda > 0\}$.

Now let us consider in $L^2(\mathbb{R})$ the perturbed operator with smooth function $b(x)$

$$\tilde{A}v(x) = v''(x) + b(x)v(x)$$

and its approximation, for example,

$$(\tilde{A}_n v_n)(x) = \frac{v_n(x+h) - 2v_n(x) + v_n(x-h)}{h^2} + b(x)v_n(x);$$

for simplicity, impose the condition $b(x) \to b$ as $x \to \pm\infty$. The operator

$$((\tilde{A} - A)v)(x) = (b(x) - b)v(x)$$

is an additive perturbation. We assume that C_0-semigroup $e^{t\tilde{A}}$, $t \in \mathbb{R}_+$, is hyperbolic. The perturbation $\tilde{A} - A$ is a relatively compact perturbation like in [107]. The operators A_n possess the same properties since $\tilde{A}_n = A_n + (\tilde{A}_n - A_n)$ and

$$e^{t_0 \tilde{A}_n} = e^{t_0 A_n} + \int_0^{t_0} A_n^\theta e^{(t_0-s)A_n} A_n^{-\theta}(\tilde{A}_n - A_n)e^{s\tilde{A}_n}\, ds. \qquad (4.10.2)$$

The crucial point is that, due to (4.10.1), such perturbations yield the estimate

$$\mu\left(\{e^{t\tilde{A}_n} x_n\}\right) \le \gamma\mu(\{x_n\}) \quad \text{with } \gamma < 1 \text{ for some } t = t_0 > 0, \qquad (4.10.3)$$

since the integral part in (4.10.2) can be estimated by any small $\epsilon > 0$ as

$$\mu\left(\int_0^{t_0-\epsilon} + \int_{t_0-\epsilon}^{t_0}\right) \le c\epsilon^{1-\theta}.$$

Then any point of the spectrum of $\tilde{A}$ located to the right of b belongs to $P\sigma(\tilde{A})$ and corresponds to a finite-dimensional generalized eigenspace. The same is valid for $\tilde{A}_n$ with $B_n = e^{t\tilde{A}_n}$, since we have (4.10.3). Using the property (4.10.3) we get from Theorem 4.6 and Proposition 4.7 the regular consistence of the operators $\lambda I_n - \tilde{A}_n$ and $\lambda I - \tilde{A}$ for any $\lambda \in i\mathbb{R}$ and any $\mathrm{Re}\,\lambda > b$.

If, as before, P is a dichotomy projector, then we have $\dim P < \infty$ and $P\tilde{A} = \tilde{A}P$. We can also state that $P_n \to P$ compactly by Proposition 4.7. This means that from the dichotomy of $\tilde{A}$ we get the dichotomy for $\tilde{A}_n$ uniformly in n by Theorem 4.14. A similar situation for concrete differential operator were considered in [256].

Chapter 5

Semilinear Fractional Equations

In the past few years, fractional differential equations have attracted the attention of many researchers due to their various applications in many fields of science: physics, mathematical biology, chemistry, nonlinear dynamics, etc. (see, e.g., $[32, 33, 87, 201, 202, 210, 212, 257]$). At the same time, there is an opinion that fractional models are more realistic and practical than the classical integer-order models (see $[34, 35, 110, 132, 151, 206, 218, 219]$). In this chapter, we discuss semilinear fractional equations in Banach spaces, the study of which has just begun (see $[36, 238, 283]$). We consider only well-posed problems. Ill-posed problems for equations with derivatives of integer order were considered, for example, in $[186, 188, 296]$. The authors intend to cover ill-posed problems for fractional equations in a separate monograph.

5.1 Setting of the Problem

In this section, we consider an approximation of the Cauchy problem

$$\left(\mathcal{D}_t^\alpha u\right)(t) = Au(t) + J^{1-\alpha} f(t, u(t)), \quad 0 \le t \le T; \quad u(0) = u^0, \quad (5.1.1)$$

where $\mathcal{D}_t^\alpha$ is the Caputo–Dzhrbashyan derivative, the operator A generates an analytic and compact resolution family $S_\alpha(\cdot, A)$, and a function $f(\cdot, \cdot)$ is sufficiently smooth.

Recall some definitions (see $[151]$). The fractional integral of order $\alpha > 0$ is defined by the formula

$$\left(J^\alpha q\right)(t) := (g_\alpha * q)(t), \quad t > 0,$$

where

$$g_\alpha(t) := \begin{cases} \dfrac{t^{\alpha-1}}{\Gamma(\alpha)}, & t > 0, \\ 0, & t \le 0, \end{cases}$$

and $\Gamma(\alpha)$ is the gamma function. The Riemann–Liouville derivative of order $\alpha > 0$ is defined by the formula

$$\left(D_t^{\alpha} q\right)(t) = \left(\frac{d}{dt}\right)^m \left(J^{m-\alpha} q\right)(t),$$

where $m = \lceil \alpha \rceil$, and the Caputo–Dzhrbashyan fractional derivative of order $\alpha > 0$ is defined by the formula

$$\left(\mathcal{D}_t^{\alpha} q\right)(t) = \left(D_t^{\alpha} q\right)(t) - \sum_{k=0}^{m-1} \frac{q^{(k)}(0)}{\Gamma(k-\alpha+1)} t^{k-\alpha}.$$

Definition 5.1. A family $\{S_{\alpha}(t, A)\}_{t\geq 0} \subset B(E)$ is called an α-*times resolution family generated by* A if the following conditions are satisfied:

(a) $S_{\alpha}(t, A)$ is strongly continuous for $t \geq 0$ and $S_{\alpha}(0, A) = I$;

(b) $S_{\alpha}(t, A)D(A) \subseteq D(A)$ and $AS_{\alpha}(t, A)x = S_{\alpha}(t, A)Ax$ for all $x \in D(A)$ and $t \geq 0$;

(c) for $x \in D(A)$, $S_{\alpha}(t, A)x$ satisfies the resolvent equation

$$S_{\alpha}(t, A)x = x + \int_0^t g_{\alpha}(t-s)S_{\alpha}(s, A)Ax\,ds, \quad t \geq 0. \qquad (5.1.2)$$

Definition 5.2. An α-times resolution family $S_{\alpha}(\cdot, A)$ is said to be *analytic* if $S_{\alpha}(\cdot, A)$ admits an analytic extension to the sector $\Sigma_{\theta_0} \setminus \{0\}$ for some $\theta_0 \in (0, \pi/2]$, where $\Sigma_{\theta_0} := \{\lambda \in \mathbb{C} : |\arg \lambda| < \theta_0\}$. We say that an analytic resolution operator $S_{\alpha}(\cdot, A)$ has the analyticity type (θ_0, ω_0) if for each $\theta < \theta_0$ and $\omega > \omega_0$, there is $M = M(\theta, \omega)$ such that

$$\|S_{\alpha}(z, A)\| \leq Me^{\omega \operatorname{Re} z}, \quad z \in \Sigma_{\theta}.$$

Definition 5.3. An α-times resolution family $S_{\alpha}(\cdot, A)$ is called *compact* if for any $t > 0$ the operator $S_{\alpha}(t, A)$ is a compact operator.

Remark 5.1. Analytic and compact α-times resolution families $S_{\alpha}(\cdot, A)$ exist. Indeed, it is known (see [93, Chap. II, Sec. 4, 4.34. Exercises (1)]) that the operator $A = d^2/dx^2$, $D(A) = \{f \in C^2[0, 1] : f'(0) = f'(1) = 0\}$, generates a compact, analytic contraction C_0-semigroup in the space $E = C[0, 1]$. Hence, from [14] and [37, Theorem 3.3], we conclude that A generates a compact and analytic α-times resolution family $S_{\alpha}(\cdot, A)$ for any $0 < \alpha < 1$.

It was proved in [37] that the homogeneous Cauchy problem (5.1.1) is well posed if and only if A generates an α-times resolution family $S_{\alpha}(\cdot, A)$.

We assume from the beginning that the resolution family $S_\alpha(\cdot, A)$ satisfies the estimate

$$\left\| S_\alpha(t, A) \right\| \le M e^{\omega t}, \quad t \ge 0, \tag{5.1.3}$$

for some M and $\omega > 0$. In this case, for $\{\lambda^\alpha : \operatorname{Re} \lambda > \omega\} \subset \rho(A)$ we have

$$\lambda^{\alpha-1}(\lambda^\alpha I - A)^{-1} x = \int_0^\infty e^{-\lambda t} S_\alpha(t, A) x\, dt, \quad \operatorname{Re} \lambda > \omega, \quad x \in E. \tag{5.1.4}$$

In this chapter, we are interested only in well-posed Cauchy problem for differential equations of order $0 < \alpha < 1$. Note that for a bounded generator A, the family $S_\alpha(t, A)$ is given by the Mittag-Leffler function $E_\alpha(t^\alpha A)$, i.e.,

$$S_\alpha(t, A) = E_\alpha(t^\alpha A) = \sum_{j=0}^\infty \frac{(t^\alpha A)^j}{\Gamma(\alpha j + 1)}.$$

Definition 5.4. A function $u(\cdot) \in C([0, T]; E)$ is called a *mild solution* of the problem (5.1.1) if the function $u(\cdot)$ satisfies the equation

$$u(t) = S_\alpha(t, A)u^0 + \int_0^t S_\alpha(t - s, A)f(s, u(s))\, ds.$$

In [148] one can find another approach to mild solutions. In this section, we follow Definition 5.4.

Theorem 5.1 (see [96, 184]). *Let A be the generator of an α-times resolution family $S_\alpha(\cdot, A)$. Assume that a function $f(\cdot, \cdot) : [0, T] \times E \to E$ is continuous in $t \in [0, T]$ and there exists a constant $L > 0$ such that*

$$\left\| f(t, x) - f(t, y) \right\| \le L \|x - y\| \quad \text{for } t \in [0, T], \ x, y \in E.$$

Then there exists a unique mild solution $u(\cdot) \in C([0, T]; E)$ of the problem (5.1.1).

We note that approximations of linear fractional problems

$$(\mathcal{D}_t^\alpha u)(t) = Au(t) + f(t), \quad u(0) = u^0,$$

were investigated mostly in Hilbert spaces in many papers (see [5, 144, 191]). Semidiscrete approximation of fractional equations were considered in [103, 135, 136].

The semidiscrete approximation of the problem (5.1.1) on the general discretization scheme is the Cauchy problems in the Banach spaces E_n

$$\begin{aligned} (\mathcal{D}_t^\alpha u_n)(t) &= A_n u_n(t) + J^{1-\alpha} f_n(t, u_n(t)), \quad 0 \le t \le T, \\ u_n(0) &= u_n^0, \end{aligned} \tag{5.1.5}$$

where operators A_n generate analytic and compact α-times resolution families $S_\alpha(\cdot, A_n)$ and functions $f_n(\cdot, \cdot)$ are sufficiently smooth. In the next section, we establish the convergence of solutions of the problems (5.1.5) to a solution of the problem (5.1.1).

5.2   Semidiscrete Approximation for a Special Case

For an analytic α-times resolution family, we get a special variant of the ABC theorem (see [195]). To prove this theorem, we need the following lemma.

Lemma 5.1 (see [195]). *Let functions $f_n(\cdot) \in C(\mathbb{R}_+; E_n)$ satisfy the estimate*

$$\|f_n(t)\|_{E_n} \leq M e^{\omega t} \quad \text{for some } M > 0, \ \omega \in \mathbb{R}, \text{ and all } n \in \mathbb{N}$$

and let $\lambda_0 \geq \omega$. The following conditions are equivalent:

  (I) *the Laplace transforms $\hat{f}_n(\cdot)$ P-converge pointwise on (λ_0, ∞) to $\hat{f}(\cdot)$ and the sequence $\{f_n(\cdot)\}$, $n \in \mathbb{N}$, is equicontinuous on compact subsets of $\mathbb{R}_+$;*
  (II) *the functions $f_n(\cdot)$ P-converge uniformly on compact subsets of $\mathbb{R}_+$ to $f(\cdot)$.*

Moreover, if the condition (II) holds, then

$$\hat{f}(\lambda) = \operatorname*{P\text{-}lim}_{n \to \infty} \hat{f}_n(\lambda)$$

for all $\lambda > \lambda_0$, where

$$f(t) := \operatorname*{P\text{-}lim}_{n \to \infty} f_n(t)$$

and $\hat{q}(\cdot)$ is the Laplace transform of $q(\cdot)$.

Theorem 5.2 (see [197]). *Assume that $0 < \alpha \leq 2$ and A and A_n generate exponentially bounded, analytic, α-times resolution families $S_\alpha(\cdot, A)$ and $S_\alpha(\cdot, A_n)$ in the Banach spaces E and E_n, respectively. The following conditions (A) and (B') are equivalent to Condition (C'):*

(A) *(consistency): there exists $\lambda \in \rho(A) \cap \bigcap\limits_{n} \rho(A_n)$ such that the resolvents converge:*

$$\left(\lambda I_n - A_n\right)^{-1} \xrightarrow{\ \mathcal{PP}\ } (\lambda I - A)^{-1};$$

(B') *(stability): there are some constants $M \geq 1$, $0 < \varphi \leq \pi/2$, and ω independent of n such that the sector $\omega + \Sigma_{\varphi + \pi/2}$ is contained in $\rho(A_n)$ and*

$$\sup_{\lambda \in \omega + \Sigma_{\beta + \pi/2}} \left\| \lambda^{\alpha-1} R\left(\lambda^\alpha; A_n\right) \right\|_{B(E_n)} \leq \frac{M}{|\lambda - \omega|} \qquad (5.2.1)$$

for any $n \in \mathbb{N}$ and any $0 < \beta < \varphi$;

(C$'$) (convergence): *for some finite $\omega_1 > 0$, we have*

$$\sup_{z \in \Sigma_\beta} e^{-\omega_1 \operatorname{Re} z} \left\| S_\alpha(z, A_n) x_n - p_n S_\alpha(z, A) x \right\|_{E_n} \to 0 \quad as \ n \to \infty$$

whenever $x_n \xrightarrow{\mathcal{P}} x$ for any $x_n \in E_n$, $x \in E$, and any $0 < \beta < \varphi$.

Proof. First, assume that Conditions (A) and (B$'$) hold. Recall (see [37]) that Condition (B$'$) is equivalent to the following condition: there are some constants C, ω, and $0 < \varphi \leq \pi/2$ such that

$$\left\| S_\alpha(z, A_n) \right\|_{B(E_n)} \leq C e^{\omega \operatorname{Re} z}$$

for any $z \in \Sigma(\beta)$, $0 < \beta < \varphi$, uniformly in $n \in \mathbb{N}$. Moreover, we can write (see [37, 241])

$$S_\alpha(z, A_n) = \frac{1}{2\pi i} \int_\Gamma e^{\lambda z} \lambda^{\alpha-1} R(\lambda^\alpha; A_n) \, d\lambda, \tag{5.2.2}$$

where Γ is a positively oriented path, which is the boundary of $\omega + \Sigma_{\beta+\pi/2}$. Partitioning the contour $\Gamma = \Gamma_a \cup \Gamma_b$, where

$$\Gamma_a = \Gamma \cap \{z : |z| \leq a\}, \quad \Gamma_b = \Gamma \setminus \Gamma_a,$$

one can make the integral over Γ_b less than $\varepsilon > 0$ uniformly in n if a is sufficiently large. For any $r > 0$, it is possible to find a sufficiently small number δ such that

$$\left\| \int_{\Gamma_a} e^{\lambda z_1} \lambda^{\alpha-1} R(\lambda^\alpha; A_n) \, d\lambda - \int_{\Gamma_a} e^{\lambda z_2} \lambda^{\alpha-1} R(\lambda^\alpha; A_n) \, d\lambda \right\|_{B(E_n)}$$

$$= \int_{\Gamma_a} \left| e^{\lambda z_1} - e^{\lambda z_2} \right| \cdot \left\| \lambda^{\alpha-1} R(\lambda^\alpha; A_n) \right\|_{B(E_n)} \, d\lambda < \varepsilon$$

when $z_1, z_2 \in \Sigma(\beta) \cap \{z : \operatorname{Re} z \leq r\} \setminus \{0\}$ and $|z_1 - z_2| < \delta$. Thus, we get the equicontinuity of $\{S_\alpha(z, A_n)\}$ on $\Sigma(\beta) \cap \{z : \operatorname{Re} z \leq r\} \setminus \{0\}$. Actually, we have proved the equicontinuity of $\{S_\alpha(z, A_n) x_n\}$ at $z = 0$ (see [195, Theorem 7]); then $\{S_\alpha(z, A_n) x_n\}$ is equicontinuous on $\Sigma(\beta) \cap \{z : \operatorname{Re} z \leq r\}$. Then it follows from Lemma 5.1 that

$$\max_{z \in \Sigma_\beta \cap \{z : \operatorname{Re} z \leq r\}} \left\| S_\alpha(z, A_n) x_n - p_n S_\alpha(z, A) x \right\|_{E_n} \to 0 \quad \text{for every } r > 0.$$

On the other hand, for any $\varepsilon > 0$ and chosen $\omega_1 > \omega$, there exists $r_0 > 0$ such that

$$\max_{z \in \Sigma_\beta \cap \{z : \operatorname{Re} z > r_0\}} e^{-\omega_1 \operatorname{Re} z} \left\| S_\alpha(z, A_n) x_n - p_n S_\alpha(z, A) x \right\|_{E_n} < \varepsilon, \quad n \geq n_0.$$

Hence we obtain Condition (C$'$).

If Condition (C′) holds, we can get Condition (A) from (5.1.4) by the Lebesgue dominated convergence theorem. From (C′) it follows that

$$\max_{z \in \Sigma(\beta)} e^{-\omega \operatorname{Re} z} \big\| S_\alpha(z, A_n) \big\|_{B(E_n)} \le C,$$

which is equivalent to Condition (B′). On the contrary, assume that it is invalid. Then one can find sequences $\|x_n\|_{E_n} = 1$ and $z_n \in \Sigma(\beta)$ such that

$$e^{-\omega \operatorname{Re} z_n} \big\| S_\alpha(z_n, A_n) x_n \big\|_{E_n} \to \infty.$$

In this case, for the sequence

$$y_n = \frac{x_n}{e^{-\omega \operatorname{Re} z_n} \big\| S_\alpha(z_n, A_n) x_n \big\|_{E_n}},$$

which converges to zero, we have

$$e^{-\omega \operatorname{Re} z_n} S_\alpha(z_n, A_n) y_n \to e^{-\omega \operatorname{Re} z_n} S_\alpha(z_n, A_n) 0 = 0$$

uniformly with respect to $z_n \in \Sigma(\beta)$. This contradicts the relation $e^{-\omega \operatorname{Re} z_n} \big\| S_\alpha(z_n, A_n) y_n \big\|_{E_n} = 1$. $\square$

Remark 5.2. From the proof of Theorem 5.2 we get the equicontinuity of $S_\alpha(\cdot, A_n)$ on any compact subset of $\overline{\mathbb{R}}^+$ if Conditions (A) and (B′) hold.

Remark 5.3. We can also get the equicontinuity of $\{R(\lambda^\alpha; A_n)\}$ on any compact subset of $\omega + \Sigma_{\beta + \pi/2}$ from Condition (B′). Actually, such equicontinuity follows from the Hilbert identity

$$\big(\lambda^\alpha I_n - A_n\big)^{-1} - \big(\mu^\alpha I_n - A_n\big)^{-1} = \big(\mu^\alpha - \lambda^\alpha\big)\big(\lambda^\alpha I_n - A_n\big)^{-1}\big(\mu^\alpha I_n - A_n\big)^{-1}$$

and the inequality (5.2.1).

Theorem 5.3 (see [196]). *Let the operator A be the generator of an exponentially bounded, analytic, α-times resolution family $S_\alpha(\cdot, A)$ in a Banach space E. Then the compactness of $S_\alpha(t, A)$ for any $t > 0$ is equivalent to the compactness of $R(\lambda^\alpha; A)$ for any $\lambda^\alpha \in \rho(A)$.*

Proof. Assume that the function $S_\alpha(\cdot, A)$ is analytic and compact. Then due to the representation

$$R(\lambda^\alpha; A) = \lambda^{1-\alpha} \int_0^\infty e^{-\lambda t} S_\alpha(t, A) dt$$

we conclude that the operator-function

$$R_{q,Q}(\lambda) := \lambda^{1-\alpha} \int_q^Q e^{-\lambda t} S_\alpha(t, A) \, dt$$

approximates the resolvent:

$$\|R(\lambda^\alpha; A) - R_{q,Q}(\lambda)\| \to 0 \quad \text{as } q \to 0 \text{ and } Q \to \infty.$$

The operator-function $R_{q,Q}(\lambda)$ is compact by [300]. Hence $R(\lambda^\alpha; A)$ is compact as a uniform limit of compact operators.

Conversely, using (5.2.2), we can write

$$S_\alpha(t, A) = \frac{1}{2\pi i} \int_\Gamma e^{\lambda t} \lambda^{\alpha-1} R(\lambda^\alpha; A) \, d\lambda \quad \text{for any } t > 0.$$

Since $R(\lambda^\alpha; A)$ is compact for any $\lambda^\alpha \in \rho(A)$, we conclude that $S_\alpha(t, A)$ is also compact for any $t \neq 0$. □

Theorem 5.4 (see [196]). *Assume that $S_\alpha(\cdot, A_n)$ and $S_\alpha(\cdot, A)$ are the analytic α-times resolution families generated by A_n and A in the Banach spaces E_n and E, respectively. Assume that Conditions* (A) *and* (B′) *hold and the resolvents $R(\lambda^\alpha; A_n)$ and $R(\lambda^\alpha; A)$ are compact. Then the following conditions are equivalent:*

(1) $R(\lambda^\alpha; A_n) \xrightarrow{\mathcal{PP}} R(\lambda^\alpha; A)$ *compactly for some $\lambda^\alpha \in \rho(A) \cap \bigcap_n \rho(A_n)$;*

(2) $S_\alpha(t, A_n) \xrightarrow{\mathcal{PP}} S_\alpha(t, A)$ *compactly for any $t > 0$.*

Proof. (1)⇒(2). Theorem 5.2 implies that $S_\alpha(t, A_n) \xrightarrow{\mathcal{PP}} S_\alpha(t, A)$. Then we only need to show that $\{S_\alpha(t, A_n)x_n\}$ is P-compact for any $\{x_n\}$ with $\|x_n\|_{E_n} = O(1)$. Let $\mu(\cdot)$ be a discrete noncompactness measure of sequences. For any $t > 0$, using (5.2.2), we get

$$\mu(\{S_\alpha(t, A_n)x_n\}) = \mu\left(\left\{\frac{1}{2\pi i} \int_\Gamma e^{\lambda t} \lambda^{\alpha-1} R(\lambda^\alpha; A_n)x_n], d\lambda\right\}\right)$$

$$= \mu\left(\left\{\frac{1}{2\pi i} \int_{\Gamma_a} e^{\lambda t} \lambda^{\alpha-1} R(\lambda^\alpha; A_n)x_n \, d\lambda\right\}\right)$$

$$+ \mu\left(\left\{\frac{1}{2\pi i} \int_{\Gamma_b} e^{\lambda t} \lambda^{\alpha-1} R(\lambda^\alpha; A_n)x_n \, d\lambda\right\}\right),$$

where Γ, Γ_a, and Γ_b are chosen as in Theorem 5.2. We can make the second term less than $\varepsilon > 0$. The first term is equal to zero due to the equicontinuity of $\{R(\lambda^\alpha; A_n)x_n\}$ on Γ_a (see Remark 5.3). Hence we conclude that $\{S_\alpha(t, A_n)x_n\}$ is P-compact.

(2)⇒(1). Conversely, since $S_\alpha(t, A_n) \xrightarrow{\mathcal{PP}} S_\alpha(t, A)$ compactly for any $t > 0$, we conclude that $S_\alpha(t, A_n) \xrightarrow{\mathcal{PP}} S_\alpha(t, A)$ and $\{S_\alpha(t, A_n)x_n\}$ is P-compact for any $\{x_n\}$ with $\|x_n\|_{E_n} = O(1)$ and any $t > 0$. Obviously, we have $R(\lambda^\alpha; A_n) \xrightarrow{\mathcal{PP}} R(\lambda^\alpha; A)$. Actually, we only need to show that $\mu(\{R(\lambda^\alpha; A_n)x_n\}) = 0$ for any $\{x_n\}$ with $\|x_n\|_{E_n} = O(1)$. We have

$$
\mu(\{R(\lambda^\alpha; A_n)x_n\}) = \mu\left(\left\{\lambda^{1-\alpha}\int_0^\infty e^{-\lambda t}S_\alpha(t, A_n)x_n\, dt\right\}\right)
$$
$$
\leq \mu\left(\left\{\lambda^{1-\alpha}\int_0^q e^{-\lambda t}S_\alpha(t, A_n)x_n\, dt\right\}\right)
$$
$$
+ \mu\left(\left\{\lambda^{1-\alpha}\int_Q^\infty e^{-\lambda t}S_\alpha(t, A_n)x_n\, dt\right\}\right)
$$
$$
+ \mu\left(\left\{\lambda^{1-\alpha}\int_q^Q e^{-\lambda t}S_\alpha(t, A_n)x_n\, dt\right\}\right),
$$

where q is small and Q is sufficiently large. The first and second terms are less than ε. From Remark 5.2, we know that $\{S_\alpha(t, A_n)\}$ is equicontinuous on $[q, Q]$. Then the third term is equal to zero. The proof is complete. $\square$

Theorem 5.5 (see [196]). *Let A and A_n be generators of analytic α-times resolution families $S_\alpha(\cdot, A)$ and $S_\alpha(\cdot, A_n)$, respectively. Assume that Conditions (A) and (B$'$) hold, the compact resolvents $R(\lambda^\alpha; A_n)$ and $R(\lambda^\alpha; A)$ converge compactly, i.e., $R(\lambda^\alpha; A_n) \xrightarrow{\mathcal{PP}} R(\lambda^\alpha; A)$ compactly for some $\lambda^\alpha \in \rho(A)$ and $u_n^0 \xrightarrow{\mathcal{P}} u^0$. Assume also that the following conditions are fulfilled:*

(i) *the functions f_n and f are continuous in both arguments and there exists a constant $\bar{M}$ independent of n such that*

$$
\sup_{\substack{t\in[0,T],\\ \|x_n\|_{E_n}\leq 1}} \|f_n(t, x_n)\|_{E_n} \leq \bar{M};
$$

(ii) *$f(\cdot, \cdot)$ is such that there exists a unique mild solution $u^*(t)$ of (5.1.1) on $[0, T]$ (for example, $f(\cdot, \cdot)$ satisfies the condition of Theorem 5.2);*

(iii) *$f_n(t, x_n) \xrightarrow{\mathcal{P}} f(t, x)$ uniformly in $t \in [0, T]$ for $x_n \xrightarrow{\mathcal{P}} x$.*

Then for almost all numbers n, the problems (5.1.5) have mild solutions $u_n^*(t)$, $t \in [0, T]$, in a neighborhood of $p_n u^*(t)$. Each sequence $\{u_n^*(t)\}$ is $\mathcal{P}$-compact and $u_n^*(t) \xrightarrow{\mathcal{P}} u^*(t)$ uniformly in $t \in [0, T]$.

Proof. From Theorem 5.4 we obtain that $S_\alpha(t, A_n) \xrightarrow{\mathcal{PP}} S_\alpha(t, A)$ compactly for any $t > 0$. Actually, we know that

$$
u(t) = S_\alpha(t, A)u^0 + \int_0^t S_\alpha(t - s, A)f\big(s, u(s)\big)\, ds,
$$

$$u_n(t) = S_\alpha(t, A_n)u_n^0 + \int_0^t S_\alpha(t - s, A_n)f_n\big(s, u_n(s)\big)\, ds$$

are mild solutions of the problems (5.1.1) and (5.1.5), respectively. We set

$$(Ku)(t) := S_\alpha(t, A)u^0 + \int_0^t S_\alpha(t - s, A)f\big(s, u(s)\big)\, ds,$$

$$(K_n u_n)(t) := S_\alpha(t, A_n)u_n^0 + \int_0^t S_\alpha(t - s, A_n)f_n\big(s, u_n(s)\big)\, ds.$$

First, we show that the operator $K : C([0, T]; E) \to C([0, T]; E)$ is compact. Let $\{u_k(\cdot)\}$ be a set of functions $u_k(\cdot) \in C([0, T]; E)$ such that $\sup_{t \in [0,T]} \|u_k(t)\| = O(1)$. From (i) we know that $f(\cdot, \cdot)$ is bounded, and then we conclude that the set of the functions $\{(Ku_k)(\cdot)\}$ is uniformly bounded, where $(Ku_k)(\cdot) \in C([0, T]; E)$. For $0 < t_1 < t_2 \le T$,

$$\big\|(Ku_k)(t_2) - (Ku_k)(t_1)\big\| = \Big\| S_\alpha(t_2, A)u_k^0 - S_\alpha(t_1, A)u_k^0$$

$$+ \int_0^{t_2} S_\alpha(t_2 - s, A)f\big(s, u_k(s)\big)\, ds$$

$$- \int_0^{t_1} S_\alpha(t_1 - s, A)f\big(s, u_k(s)\big)ds \Big\|$$

$$\le \big\| S_\alpha(t_2, A)u_k^0 - S_\alpha(t_1, A)u_k^0 \big\|$$

$$+ \int_0^{t_1 - \delta} \big\| S_\alpha(t_2 - s, A) - S_\alpha(t_1 - s, A) \big\| \, \big\| f\big(s, u_k(s)\big) \big\|\, ds$$

$$+ \int_{t_1 - \delta}^{t_1} \big\| S_\alpha(t_2 - s, A) - S_\alpha(t_1 - s, A) \big\| \, \big\| f\big(s, u_k(s)\big) \big\|\, ds$$

$$+ \int_{t_1}^{t_2} \big\| S_\alpha(t_2 - s, A) \big\| \, \big\| f\big(s, u_k(s)\big) \big\|\, ds$$

$$\le \big\| S_\alpha(t_2, A) - S_\alpha(t_1, A) \big\|$$

$$+ \bar{M}(t_1 - \delta) \sup_{s \in [0, t_1 - \delta]} \big\| S_\alpha(t_2 - s, A) - S_\alpha(t_1 - s, A) \big\|$$

$$+ 2M\bar{M}\delta + M\bar{M}(t_2 - t_1) \to 0 \quad \text{as } |t_2 - t_1| \to 0.$$

This means that $\{(Ku_k)(\cdot)\}$ is equicontinuous. Now we will prove that the sequence $\{(Ku_k)(t)\}$ is relatively compact for any $t > 0$. From Theorem 5.3, we see that $S_\alpha(t, A)$ is compact for any $t > 0$. Choosing $0 < \varepsilon < t$

arbitrarily, we conclude that $S_\alpha(\varepsilon, A)$ is compact,

$$\left\{ S_\alpha(\varepsilon, A) \int_0^{t-\varepsilon} S_\alpha(t - s - \varepsilon, A) f(s, u_k(s)) \, ds \right\}$$

is relatively compact for any $t > 0$. Now

$$\left\| S_\alpha(\varepsilon, A) \int_0^{t-\varepsilon} S_\alpha(t - s - \varepsilon, A) f(s, u_k(s)) \, ds \right.$$

$$\left. - \int_0^{t-\varepsilon} S_\alpha(t - s, A) f(s, u_k(s)) \, ds \right\|$$

$$\leq \int_0^{t-\varepsilon} \left\| S_\alpha(\varepsilon, A) S_\alpha(t - s - \varepsilon, A) - S_\alpha(t - s, A) \right\| \left\| f(s, u_k(s)) \right\| ds$$

$$\leq \int_0^{t-\varepsilon-\delta} \left\| S_\alpha(\varepsilon, A) S_\alpha(t - s - \varepsilon, A) - S_\alpha(t - s, A) \right\| \left\| f(s, u_k(s)) \right\| ds$$

$$+ \int_{t-\varepsilon-\delta}^{t-\varepsilon} \left\| S_\alpha(\varepsilon, A) S_\alpha(t - s - \varepsilon, A) - S_\alpha(t - s, A) \right\| \left\| f(s, u_k(s)) \right\| ds$$

$$\leq \bar{M} \int_0^{t-\varepsilon-\delta} \left\| S_\alpha(\varepsilon, A) S_\alpha(t - s - \varepsilon, A) - S_\alpha(t - s, A) \right\| ds$$

$$+ (M^2 + M) \bar{M} \int_{t-\varepsilon-\delta}^{t-\varepsilon} ds \to 0$$

as $\delta \to 0$ and $\varepsilon \to 0$. We get the limit of the first term from [96, Lemma 3.4]. Moreover, for $\varepsilon \to 0$ we have

$$\left\| S_\alpha(\varepsilon, A) \int_0^{t-\varepsilon} S_\alpha(t - s - \varepsilon, A) f(s, u_k(s)) \, ds \right.$$

$$\left. - \int_0^t S_\alpha(t - s, A) f(s, u_k(s)) \, ds \right\|$$

$$\leq \left\| S_\alpha(\varepsilon, A) \int_0^{t-\varepsilon} S_\alpha(t - s - \varepsilon, A) f(s, u_k(s)) \, ds \right.$$

$$\left. - \int_0^{t-\varepsilon} S_\alpha(t - s, A) f(s, u_k(s)) \, ds \right\|$$

$$+ \left\| \int_{t-\varepsilon}^t S_\alpha(t - s, A) f(s, u_k(s)) \, ds \right\|$$

$$\leq \left\| S_\alpha(\varepsilon, A) \int_0^{t-\varepsilon} S_\alpha(t - s - \varepsilon, A) f\big(s, u_k(s)\big) \, ds \right.$$

$$- \int_0^{t-\varepsilon} S_\alpha(t - s, A) f\big(s, u_k(s)\big) \, ds \left.\right\|$$

$$+ \int_{t-\varepsilon}^t M \bar{M} ds \to 0.$$

This means that

$$\left\{ \int_0^t S_\alpha(t - s, A) f\big(s, u_k(s)\big) \, ds \right\}$$

is relatively compact for any $t > 0$. Obviously, when $t = 0$, it is also relatively compact. This means that for any t, $\{(K u_k)(t)\}$ is a relatively compact set in E. Then the generalized Arzelà–Ascoli theorem implies that $\{(K u_k)(t)\}$ is a compact set in E and hence $\{(K u_k)(\cdot)\}$ is compact in $C([0, T]; E)$. Hence, K is a compact operator. Similarly, K_n are also compact.

If $u_n(\cdot) \xrightarrow{\mathcal{P}} u(\cdot)$, then the condition (iii) implies that

$$f_n(t, u_n(t)) \xrightarrow{\mathcal{P}} f(t, u(t)) \quad \text{uniformly in } t \in [0, T].$$

We have

$$\sup_{t \in [0,T]} \big\| (K_n u_n)(t) - p_n(K u)(t) \big\|_{E_n}$$

$$\leq \sup_{t \in [0,T]} \big\| S_\alpha(t, A_n) u_n^0 - p_n S_\alpha(t, A) u^0 \big\|_{E_n}$$

$$+ \sup_{t \in [0,T]} \left\| \int_0^t S_\alpha(t - s, A_n) f_n\big(s, u_n(s)\big) \, ds \right.$$

$$- p_n \int_0^t S_\alpha(t - s, A) f\big(s, u(s)\big) \, ds \left.\right\|_{E_n} \to 0.$$

Hence we get $K_n \xrightarrow{\mathcal{PP}} K$. Let $\{u_n(\cdot)\}$ be any sequence of functions $u_n(\cdot) \in C([0, T]; E_n)$ such that $\sup_{t \in [0,T]} \|u_n(t)\|_{E_n} = O(1)$. Next we show that $\mu(\{(K_n u_n)(t)\}) = 0$ for all $t \in [0, T]$. Indeed,

$$\mu\big(\{(K_n u_n)(t)\}\big) \leq \mu\big(\{S_\alpha(t, A_n) u_n^0\}\big)$$

$$+ \mu \left(\left\{ \int_0^t S_\alpha(t - s, A_n) f_n\big(s, u_n(s)\big) \, ds \right\} \right)$$

$$\leq \mu\big(\{S_\alpha(t, A_n)u_n^0\}\big)$$

$$+ \mu\left(\left\{\int_0^{t-\delta} S_\alpha(t - s, A_n)f_n\big(s, u_n(s)\big)\, ds\right\}\right)$$

$$+ \sup_{t\in[0,T]} \left\| \int_{t-\delta}^t S_\alpha(t - s, A_n)f_n\big(s, u_n(s)\big) ds \right\|_{E_n}.$$

Obviously, the first term is equal to zero. Since

$$\big\|S_\alpha(t, A_n)\big\|_{B(E_n)} \leq M e^{\omega t}, \qquad \sup_{t\in[0,T]} \big\|f_n(t, u_n(t))\big\|_{E_n} \leq \bar{M},$$

we see that the third term can be made less than ε by choosing sufficient small δ. Lemma 5.1 and the condition (iii) imply that $\{f_n(s, u_n(s))\}$ is equicontinuous on $[0, t - \delta]$. We know that $\{S_\alpha(t - s, A_n)\}$ is also equicontinuous on $[0, t - \delta]$. Thus,

$$\mu\left(\left\{\int_0^{t-\delta} S_\alpha(t - s, A_n)f_n\big(s, u_n(s)\big) ds\right\}\right) = 0.$$

Hence, $\{(K_n u_n)(\cdot)\}$ is P-compact. So we conclude that $K_n \xrightarrow{\mathcal{PP}} K$ compactly.

It follows from the assumption that K has no fixed point on the boundary ∂S_r, where

$$S_r = \left\{u: \sup_{t\in[0,T]} \big\|u(t) - u^*(t)\big\| < r\right\}.$$

The results of [288] imply that $\gamma(I - K; \partial S_r) = \gamma(I_n - K_n; \partial S_{n,r})$ for $n \geq n_0$. If we verify that $\gamma(I - K; \partial S_r) \neq 0$, then from [288, Theorem 3] we conclude that solutions of the problem (5.1.5) exist in a neighborhood of $p_n u^*(t)$, each sequence $\{u_n^*(t)\}$ is P-compact, and $u_n^*(t) \xrightarrow{P} u^*(t)$ uniformly with respect to $t \in [0, T]$. Then we obtain our conclusion.

Since the family $S_\alpha(\cdot, A)$ is analytic and compact, the proof of the fact that $\gamma(I - K; \partial S_r) = 1$ literally repeats the proof of [295, Theorem 14.2.3]. Theorem 5.5 is proved. □

5.3   The Case of a Classical Right-Hand Side

In this section, we consider the approximation of the well-posed Cauchy problem

$$\big(\mathcal{D}_t^\alpha u\big)(t) = Au(t) + f(t, u(t)), \quad 0 < t \leq T; \quad u(0) = u^0, \qquad (5.3.1)$$

where $\mathcal{D}_t^\alpha$ is the Caputo–Dzhrbashyan derivative, the operator A generates an analytic and compact α-times resolvent family $S_\alpha(\cdot, A)$, and the function $f(\cdot, \cdot)$ is sufficiently smooth in both arguments.

Definition 5.5 (see [148]). A family $\{P_\alpha(t, A)\}_{t\geq 0}$ of strongly continuous functions $P_\alpha(\cdot, A) : (0, \infty) \to B(E)$ is called an (α, α)-*times resolvent family generated by* A if there exists $\omega \geq 0$ such that $\{\lambda^\alpha : \operatorname{Re}\lambda > \omega\} \subset \rho(A)$ and

$$\left(\lambda^\alpha I - A\right)^{-1} x = \int_0^\infty e^{-\lambda t} P_\alpha(t, A) x \, dt, \quad \operatorname{Re}\lambda > \omega, \quad x \in E. \quad (5.3.2)$$

Remark 5.4 (see [37, 148, 182]). Obviously, for $1 < \alpha < 2$, if A generates an α-times resolvent family $S_\alpha(t, A)$, then it is also the generator of an (α, α)-times resolvent family $P_\alpha(t, A)$ and

$$P_\alpha(t, A) = \left(g_{\alpha-1} * S_\alpha\right)(t). \quad (5.3.3)$$

While, when $0 < \alpha < 1$, if A generates an analytic α-times resolvent family $S_\alpha(t, A)$, then it is also the generator of an analytic (α, α)-times resolvent family

$$P_\alpha(t, A) = \frac{1}{2\pi i} \int_\Gamma e^{\lambda t} R\left(\lambda^\alpha; A\right) d\lambda$$

and

$$\left(g_{1-\alpha} * P_\alpha\right)(t) = S_\alpha(t, A). \quad (5.3.4)$$

For $P_\alpha(\cdot, A)$, we have the following properties (see [148, 182]):

$$P_\alpha(t, A)x = g_\alpha(t)x + A \int_0^t g_\alpha(t - s) P_\alpha(s, A) x \, ds, \quad \text{for any } x \in E, \, t > 0,$$

$$AP_\alpha(t, A)x = P_\alpha(t, A)Ax \qquad\qquad \text{for any } x \in D(A).$$

For $0 < \alpha < 1$, the following lemma holds.

Lemma 5.2 (see [182]). *Let A be the generator of an analytic α-times resolution family $S_\alpha(t, A)$. The following assertions hold:*

(1) $P_\alpha(t, A) \in B(E)$ *and*

$$\left\|P_\alpha(t, A)\right\| \leq M e^{\omega t}(1 + t^{\alpha-1}) \quad \text{for any } t > 0;$$

(2) *for any $x \in E$, $P_\alpha(t, A)x \in D(A)$ and*

$$\left\|AP_\alpha(t, A)\right\| \leq M e^{\omega t}(1 + t^{-1}) \quad \text{for any } t > 0;$$

(3) $S'_\alpha(t, A) = AP_\alpha(t, A)$ for any $t > 0$, $R(P_\alpha^{(l)}(t, A)) \subseteq D(A)$ for any integer $l \geq 0$, and

$$\left\|A^k P_\alpha^{(l)}(t, A)\right\| \leq Me^{\omega t}\left(1 + t^{-l-1-\alpha(k-1)}\right) \quad \text{for any } t > 0,$$

where $k = 0, 1$.

Definition 5.6. A function $u(\cdot) \in C([0, T]; E)$ is called a *mild solution* of the problem (5.3.1) if it satisfies the equation

$$u(t) = S_\alpha(t, A)u^0 + \int_0^t P_\alpha(t - s, A)f\big(s, u(s)\big)\,ds. \tag{5.3.5}$$

Theorem 5.6 (see [96, 184, 199, 237]). *Let A be the generator of an analytic α-times resolution family $S_\alpha(\cdot, A)$ on a Banach space E. Assume that the function $f(\cdot, \cdot) : [0, T] \times E \to E$ is continuous in both arguments and there exists a constant $L > 0$ such that*

$$\left\|f(t, x) - f(t, y)\right\| \leq L\|x - y\| \quad \text{for } t \in [0, T] \text{ and } x, y \in E.$$

Then there exists a unique mild solution $u(\cdot) \in C([0, T]; E)$ of the problem (5.3.1). Moreover, the mapping $u(0) \to u(\cdot)$ from E into $C([0, T]; E)$ is Lipschitz continuous.

Proof. We consider operator

$$K : C([0, T]; E) \to C([0, T]; E),$$

$$(Ku)(t) := S_\alpha(t, A)u^0 + \int_0^t P_\alpha(t - s, A)f\big(s, u(s)\big)\,ds, \quad t \in [0, T].$$

In order to prove that the problem (5.3.1) has a unique mild solution, we need to show that the operator K has a fixed point. Let $u, v \in C([0, T]; E)$; then from the condition on the function $f(\cdot, \cdot)$ we conclude that

$$\left\|(Ku)(t) - (Kv)(t)\right\| \leq \int_0^t \left\|P_\alpha(t - s, A)\right\| \cdot \left\|f(s, u(s)) - f(s, v(s))\right\|\,ds$$

$$\leq ML \int_0^t g_\alpha(t - s)\left\|u(s) - v(s)\right\|ds$$

$$\leq MLt^\alpha\|u - v\|_{C([0, T]; E)}.$$

Thus,

$$\sup_{t \in [0, T]} \left\|(K^n u)(t) - (K^n v)(t)\right\| \leq \sup_{t \in [0, T]} \frac{(MLt^\alpha)^n}{n!}\|u - v\|_{C([0, T]; E)}$$

$$\leq \frac{(MLT^\alpha)^n}{n!}\|u - v\|_{C([0, T]; E)}.$$

Since $(MLT^\alpha)^n/n! < 1$ for sufficiently large n, K is a contraction. Hence due to the contraction mapping theorem we conclude that K has a unique fixed point which is a solution of (5.3.1).

Let $v(\cdot)$ be a mild solution of (5.3.1). Then the following inequalities are valid:

$$\left\|u(t) - v(t)\right\| \leq \left\|S_\alpha(t, A)u(0) - S_\alpha(t, A)v(0)\right\|$$

$$+ \int_0^t \left\|P_\alpha(t - s, A)\right\| \cdot \left\|f(s, u(s)) - f(s, v(s))\right\| ds$$

$$\leq M\|u^0 - v^0\| + ML \int_0^t g_\alpha(t - s)\|u(s) - v(s)\| ds.$$

By the Gronwall inequality,

$$\left\|u(t) - v(t)\right\| \leq Me^{MLT^\alpha}\|u(0) - v(0)\|.$$

This means that the solution $u(\cdot)$ is unique and the mapping $u(0) \to u(\cdot)$ is Lipschitz continuous. $\qquad\square$

Note that problems similar to (5.3.1) and $(*)$ on Chapter 1 in the case where $\exp(\cdot A)$ is not a compact but condensing semigroup were considered in $[20, 33, 66, 139, 215, 216]$.

5.4 Semidiscrete Approximation for the Classical Case

The semidiscrete approximation on the general discretization scheme of the problem (5.3.1) consists of the following Cauchy problems in the Banach spaces E_n:

$$\left(\mathcal{D}_t^\alpha u_n\right)(t) = A_n u_n(t) + f_n(t, u_n(t)), \quad 0 < t \leq T, \quad u_n(0) = u_n^0, \quad (5.4.1)$$

where the operators A_n generate analytic and compact resolvent families $S_\alpha(\cdot, A_n)$ and the functions $f_n(\cdot, \cdot)$ are sufficiently smooth.

Remark 5.5. Condition (C$'$) implies that

$$\sup_{t \in [0,T]} \left\|S_\alpha(t, A_n)x_n - p_n S_\alpha(t, A)x\right\|_{E_n} \to 0 \quad \text{as } n \to \infty$$

whenever $x_n \xrightarrow{\mathcal{P}} x$ for any $x_n \in E_n$ and $x \in E$. Then Eq. (5.3.4) implies that

$$\sup_{t \in [0,T]} \left\|\int_0^t P_\alpha(t - s, A_n)x_n \, ds - p_n \int_0^t P_\alpha(t - s, A)x \, ds\right\|_{E_n}$$

$$= \sup_{t \in [0,T]} \left\|(g_1 * P_\alpha)(t, A_n)x_n - p_n(g_1 * P_\alpha)(t, A)x\right\|_{E_n}$$

$$= \sup_{t\in[0,T]} \left\| (g_\alpha * g_{1-\alpha} * P_\alpha)(t, A_n)x_n - p_n (g_\alpha * g_{1-\alpha} * P_\alpha)(t, A)x \right\|_{E_n}$$

$$= \sup_{t\in[0,T]} \left\| \int_0^t g_\alpha(t-s)S_\alpha(s, A_n)x_n \, ds - p_n \int_0^t g_\alpha(t-s)S_\alpha(s, A)x \, ds \right\|_{E_n}$$

$$= \sup_{t\in[0,T]} \left\| \int_0^t g_\alpha(t-s)\big(S_\alpha(s, A_n)x_n - p_n S_\alpha(s, A)x\big) \, ds \right\|_{E_n}$$

$$\leq \int_0^t |g_\alpha(t-s)| \, ds \sup_{t\in[0,T]} \left\| S_\alpha(s, A_n)x_n - p_n S_\alpha(s, A)x \right\|_{E_n}$$

$$= g_{\alpha+1}(t) \sup_{t\in[0,T]} \left\| S_\alpha(s, A_n)x_n - p_n S_\alpha(s, A)x \right\|_{E_n} \to 0 \quad \text{as } n \to \infty$$

whenever $x_n \xrightarrow{\mathcal{P}} x$ for any $x_n \in E_n$ and $x \in E$.

Remark 5.6. Due to the representation
$$S_\alpha(z, A_n) = \frac{1}{2\pi i} \int_\Gamma e^{\lambda z} \lambda^{\alpha-1} R(\lambda^\alpha; A_n) d\lambda,$$
where Γ is a positively oriented path, which is the boundary of $\omega + \Sigma_{\beta+\pi/2}$, we get the equicontinuity of $\{S_\alpha(z, A_n)\}$ on $\Sigma(\beta) \cap \{z : \operatorname{Re} z \leq r\} \setminus \{0\}$. Actually, we have proved the equicontinuity of the α-times resolution family $\{S_\alpha(z, A_n)\}$ at $z = 0$ (see [195, Theorem 7]), and, therefore, $\{S_\alpha(z, A_n)\}$ is equicontinuous on $\Sigma(\beta) \cap \{z : \operatorname{Re} z \leq r\}$.

Lemma 5.3. *Let $S_\alpha(t, A)$, $t \geq 0$, be an analytic α-times resolution family generated by A on a Banach space E. Then the α-times and (α, α)-times resolution families $S_\alpha(t, A)$ and $P_\alpha(t, A)$ are continuous in the uniform operator topology for all $t > 0$.*

Proof. From [96, Lemma 3.8] we know that α-times resolution family $S_\alpha(\cdot, A)$ is continuous in the uniform operator topology for all $t > 0$. From Remark 5.4 we obtain that the (α, α)-times resolution family $P_\alpha(t, A)$ generated by A is analytic. So we only need to show that $P_\alpha(t, A)$ is continuous in the uniform operator topology for all $t > 0$.

In fact, Lemma 5.2 implies that
$$\|P'_\alpha(t, A)\| \leq M(1 + t^{\alpha-2}).$$
Then for any $x \in E$, $\|x\| \leq 1$, and any $|h| < t$, we have

$$\left\| P_\alpha(t+h, A)x - P_\alpha(t, A)x \right\| = \left\| \int_t^{t+h} P'_\alpha(s, A)x \, ds \right\|$$

$$\leq M\|x\| \cdot \left\| \int_t^{t+h} s^{\alpha-2} \, ds \right\| \leq M \frac{(t+h)^{\alpha-1} - t^{\alpha-1}}{\alpha - 1} \to 0$$

as $h \to 0$. Lemma 5.3 is proved. □

Theorem 5.7 (see [199]). *Let $S_\alpha(\cdot, A)$ be an analytic α-times resolvent family generated by A in the Banach space E. Assume that $S_\alpha(t, A)$ and $P_\alpha(t, A)$ are compact for any $t > 0$. Then*

$$\lim_{h \to 0} \left\| P_\alpha(t, A) - S_\alpha(h, A) P_\alpha(t - h, A) \right\| = 0 \quad \text{for any } t > 0.$$

Proof. Lemma 5.3 implies that the α-times and (α, α)-times resolution families $S_\alpha(t, A)$ and $P_\alpha(t, A)$ are continuous in the uniform operator topology for all $t > 0$. Let $t > 0$ and $0 < h < t$. First, we prove that

$$\lim_{h \to 0} \left\| P_\alpha(t + h, A) - S_\alpha(h, A) P_\alpha(t, A) \right\| = 0 \tag{5.4.2}$$

for any $t > 0$. We set $\varepsilon > 0$. Since $P_\alpha(t, A)$ is compact, the set

$$P_t := \{ P_\alpha(t, A) x : \; \|x\| \le 1 \}$$

is also compact. Then for any $x \in E$, $\|x\| \le 1$, there exists a finite ε-net

$$\{ P_\alpha(t, A) x_1, \; P_\alpha(t, A) x_2, \; \ldots, \; P_\alpha(t, A) x_l \} \subset P_t$$

such that

$$\left\| P_\alpha(t, A) x - P_\alpha(t, A) x_i \right\| \le \frac{\varepsilon}{3(M + 1)}, \quad 1 \le i \le l. \tag{5.4.3}$$

The strong convergence $S_\alpha(h, A) x_i \to x_i$ as $h \to 0$ for all $i = 1, \ldots, l$ implies that there exists $0 < h_1 < t$ such that

$$\left\| P_\alpha(t, A) x_i - S_\alpha(h, A) P_\alpha(t, A) x_i \right\| \le \frac{\varepsilon}{3} \tag{5.4.4}$$

for all $0 \le h \le h_1$ and $1 \le i \le l$. Due to the continuity of $P_\alpha(t, A)$ in the uniform operator topology, there exists $0 < h_2 < t$ such that

$$\left\| P_\alpha(t, A) x - P_\alpha(t + h, A) x \right\| \le \frac{\varepsilon}{3} \tag{5.4.5}$$

for all $0 \le h \le h_2$ and $\|x\| \le 1$. Consequently, for any $0 \le h \le \min\{h_1, h_2\}$ and $\|x\| \le 1$, the inequalities (5.4.3), (5.4.4), and (5.4.5) imply that

$$\left\| P_\alpha(t + h, A) x - S_\alpha(h, A) P_\alpha(t, A) x \right\|$$

$$\le \left\| P_\alpha(t + h, A) x - P_\alpha(t, A) x \right\| + \left\| P_\alpha(t, A) x - P_\alpha(t, A) x_i \right\|$$

$$+ \left\| P_\alpha(t, A) x_i - S_\alpha(h, A) P_\alpha(t, A) x_i \right\|$$

$$+ \left\| S_\alpha(h, A) P_\alpha(t, A) x_i - S_\alpha(h, A) P_\alpha(t, A) x \right\|$$

$$\leq \left\| P_\alpha(t+h,A)x - P_\alpha(t,A)x \right\|$$
$$+ (M+1)\left\| P_\alpha(t,A)x - P_\alpha(t,A)x_i \right\|$$
$$+ \left\| P_\alpha(t,A)x_i - S_\alpha(h,A)P_\alpha(t,A)x_i \right\| \leq \varepsilon.$$

This means that (5.4.2) holds uniformly for any $t > 0$. On the other hand,

$$\left\| P_\alpha(t,A) - S_\alpha(h,A)P_\alpha(t-h,A) \right\|$$
$$\leq \left\| P_\alpha(t,A) - P_\alpha(t+h,A) \right\| + \left\| P_\alpha(t+h,A) - S_\alpha(h,A)P_\alpha(t,A) \right\|$$
$$+ \left\| S_\alpha(h,A)P_\alpha(t,A) - S_\alpha(h,A)P_\alpha(t-h,A) \right\|$$
$$\leq \left\| P_\alpha(t,A) - P_\alpha(t+h,A) \right\| + \left\| P_\alpha(t+h,A) - S_\alpha(h,A)P_\alpha(t,A) \right\|$$
$$+ M\left\| P_\alpha(t,A) - P_\alpha(t-h,A) \right\|.$$

Due to the continuity of (α,α)-times resolution family $\{P_\alpha(t,A)\}$ in the uniform operator topology, we have

$$\lim_{h \to 0} \left\| P_\alpha(t,A) - P_\alpha(t+h,A) \right\| = 0,$$
$$\lim_{h \to 0} \left\| P_\alpha(t,A) - P_\alpha(t-h,A) \right\| = 0.$$

Hence our conclusion follows from (5.4.2). The theorem is proved. $\square$

The first attempts to characterize the compactness of α-times resolution families by the compactness property of the corresponding α-times resolvent families was made in [14, 96]. Then a result like what follows were obtained in [196, 203].

Theorem 5.8. *Let A and A_n be generators of analytic α-times resolution families $S_\alpha(\cdot, A)$ and $S_\alpha(\cdot, A_n)$, respectively. Assume that Conditions (A) and (B$'$) are fulfilled, the compact resolvents $R(\lambda^\alpha; A_n)$ and $R(\lambda^\alpha; A)$ converge compactly, i.e., $R(\lambda^\alpha; A_n) \xrightarrow{\mathcal{PP}} R(\lambda^\alpha; A)$ compactly for some $\lambda^\alpha \in \rho(A)$, and $u_n^0 \xrightarrow{\mathcal{P}} u^0$. Assume also that the following conditions hold:*

(i) *the functions f_n and f are continuous in both arguments and there exists a constant $\bar{M}$ independent of n such that*

$$\sup_{\substack{t \in [0,T], \\ \|x_n\|_{E_n} \leq 1}} \left\| f_n(t, x_n) \right\|_{E_n} \leq \bar{M};$$

(ii) $f(\cdot,\cdot)$ *is such that there exists a unique mild solution* $u^*(t)$ *of the problem* (5.3.1) *on* $[0,T]$ (*e.g.*, $f(\cdot,\cdot)$ *satisfies the conditions of Theorem 5.6*);

(iii) $f_n(t,x_n) \xrightarrow{\mathcal{P}} f(t,x)$ *uniformly in* $t \in [0,T]$ *as* $x_n \xrightarrow{\mathcal{P}} x$.

Then the problems (5.4.1) *have mild solutions* $u_n^*(t)$, $t \in [0,T]$, *in a neighborhood of* $p_n u^*(t)$ *for almost all* n. *Each sequence* $\{u_n^*(t)\}$ *is P-compact and* $u_n^*(t) \xrightarrow{\mathcal{P}} u^*(t)$ *uniformly in* $t \in [0,T]$.

The proof of Theorem 5.8 is similar to the proof of Theorem 5.5 but is essentially based on Theorem 5.7.

5.5 Maximal Regularity in the Space $C_0^\alpha([0,T];E)$

The property of maximal regularity plays a key role in the study of various types of partial differential equations (see [1–4, 30]). This property is especially important when linearizations of nonlinear equations are examined. Consider the following Cauchy problem in an abstract Banach space E:

$$\begin{cases} u'(t) = Au(t) + f(t), & 0 \le t \le T, \\ u(0) = u^0, \end{cases} \tag{5.5.1}$$

where the operator A is a generator of a C_0-semigroup $\exp(tA)$, $t \ge 0$. There is an extensive literature devoted to the maximal regularity of the problem (5.5.1) in various spaces (see [235]). For example, in [88], Eberhardt and Greiner proved that, in general, the problem (5.5.1) does not possess the maximal regularity property in $C([0,T];E)$, except the cases where the operator A is continuous or the space E includes a subspace isomorphic to the space c_0 of infinitesimal sequences. In [30], Ashyralyev and Sobolevskii indicated that in the Hölder space $C_0^\gamma([0,T];E)$, the analyticity of a C_0-semigroup is equivalent to the maximal regularity (coercive well-posedness) of the problem (5.5.1). For more results about the maximal regularity of (5.5.1) in various spaces, we refer the reader to [30, 120, 235].

At the same time, there is an extensive literature devoted to the maximal regularity for second-order differential equations. In [27], the weak maximal regularity was considered for second-order hyperbolic differential equations in the spaces $C^\gamma([0,T];E)$, $C([0,T];E^\theta)$ and $L^p([0,T];E^\theta)$, where E^θ is an interpolation space. In [26], difference schemes for elliptic problems in $L_{\tau_n}^p([0,T];E_n)$ were presented and the corresponding discrete maximal regularity was investigated. The maximal regularity in $L^p([0,T];E)$ was also considered in [29]. For more detail, we refer the reader to [28, 235, 295, 311].

It is known that fractional differential problems have become one of the most interesting research topics due to their broad applications in physics, engineering, biology, etc. Many fruitful applications on these problems can be found, for example, in [37, 96, 99, 137, 151, 182, 185, 195, 198, 235, 236, 241, 306]. Moreover, the regularity of solutions for fractional evolution equations was also considered. For example, in [37] E. Bajlekova considered applications of fractional Cauchy problems in the theory of solution operators and examined the maximal regularity in L^p for Riemann–Liouville fractional differential equations with positive and $\mathcal{R}$-sectorial operators A. The regularity of solutions was also studied in [73, 74]. More recently, Li and Li [182] considered the Hölder regularity (not maximal regularity) for fractional Cauchy problems on $(0, 1)$ and generalized the regularity results for the first-order differential equation (5.5.1) in [225].

Also note that, by the method of operator-valued Fourier multiplier, Bu and Cai (see [61–64]) characterized the well-posedness and maximal regularity for second-order and fractional degenerate differential equations in various function spaces.

The theory of approximation for differential equations in abstract Banach spaces was developed in [4, 5, 30, 66, 69, 120, 122, 123, 167, 195–198, 235, 288, 295]. Numerical problems for homogeneous differential equations and semilinear differential equations were analyzed. Specifically, approximations for fractional differential equations in abstract spaces were investigated in [195–198] and the existence, stability, and the order of convergence of difference schemes for the general approximation of fractional Cauchy problems were obtained.

We consider the well-posedness and the maximal regularity in the Hölder space $C_0^\gamma([0, T]; E)$ and the full discrete approximation of the following nonhomogeneous fractional Cauchy problem in an abstract space E:

$$\begin{cases} D^\alpha u(t) = Au(t) + f(t), & 0 \le t \le T, \\ u(0) = u^0. \end{cases} \tag{5.5.2}$$

Here the operator A is the generator of an analytic α-resolvent family and D^α is the Caputo–Dzhrbashyan fractional derivative with $\alpha \in (0, 1)$.

5.5.1 Setting of the problem

Let $\gamma \in (0, 1)$, $T > 0$, and $C^\gamma([0, T]; E)$ be the Banach space consisting of continuous functions $u(\cdot)$ from the segment $[0, T]$ to the Banach space E

with the following norm:

$$\|u(\cdot)\|_{C^\gamma([0,T];E)} = \sup_{0 \le t \le T} \|u(t)\|_E + \sup_{0 \le t < t+\tau \le T} \frac{\|u(s+\tau) - u(s)\|_E}{\tau^\gamma}.$$

Let the Banach space $C_0^\gamma([0,T];E)$ be the completion of the space of smooth functions $u(\cdot)$ from the segment $[0,T]$ to E with the following norm:

$$\|u(\cdot)\|_{C_0^\gamma([0,T];E)} = \|u(\cdot)\|_{C([0,T];E)} + \sup_{0 \le t < t+\tau \le T} t^\gamma \frac{\|u(t+\tau) - u(t)\|_E}{\tau^\gamma}.$$

It is easy to prove that $C^\gamma([0,T];E) \subset C_0^\gamma([0,T];E) \subset C([0,T];E)$.

Definition 5.7. The problem (5.5.2) is called *well posed* in $C_0^\gamma([0,T];E)$ if the following conditions are fulfilled:

(i) for any function $f(\cdot) \in C_0^\gamma([0,T];E)$ and $u^0 \in D(A)$, there is a unique solution $u(t) = u(t; f, u^0)$ of the problem (5.5.2) on the segment $[0,T]$ and $u(\cdot) \in C_0^\gamma([0,T];E)$;

(ii) the operator $u(t) = u(t; f, u^0)$ is continuous from $C_0^\gamma([0,T];E) \times D(A)$ into $C_0^\gamma([0,T];E)$, where $C_0^\gamma([0,T];E) \times D(A)$ is equipped with the norm

$$\|(f, u^0)\|_{C_0^\gamma([0,T];E) \times D(A)} = \|f(\cdot)\|_{C_0^\gamma([0,T];E)} + \|Au^0\|_E.$$

Definition 5.8. We say that the problem (5.5.2) *satisfies the property of maximal regularity* (or the coercive well-posedness) in the space $C_0^\gamma([0,T];E)$ if it is well posed in $C_0^\gamma([0,T];E)$ and satisfies the following maximal regularity (coercive) inequality:

$$\|D^\alpha u(\cdot)\|_{C_0^\gamma([0,T];E)} + \|Au(\cdot)\|_{C_0^\gamma([0,T];E)} \le C\Big(\|f(\cdot)\|_{C_0^\gamma([0,T];E)} + \|Au^0\|_E\Big).$$

Definition 5.9. A function $u(\cdot) \in C([0,T];E)$ is called a *classical solution* of the Cauchy problem (5.5.2) if for any $t \in [0,T]$, $u(t) \in D(A)$, $Au(\cdot)$, $D^\alpha u(\cdot) \in C([0,T];E)$, and $u(\cdot)$ satisfies (5.5.2) on $[0,T]$.

We denote the set of mappings A that generate analytic α-resolvent families by $\mathcal{A}^\alpha$.

Theorem 5.9 (see [193, 200]). *Let $0 < \alpha$, $\gamma < 1$. Assume that $A \in \mathcal{A}^\alpha$, $f(\cdot) \in C_0^\gamma([0,T];E)$, and $u^0 \in D(A)$. Then there exists a unique classical solution of the problem (5.5.2), which is given by the formula*

$$u(t) = S_\alpha(t)u^0 + (P_\alpha * f)(t), \quad t \in [0,T]. \tag{5.5.3}$$

Theorem 5.10 (see [193, 200]). *Let $0 < \gamma$, $\alpha < 1$. Assume that $A \in \mathcal{A}^\alpha$, $u^0 \in D(A)$, and $f(\cdot) \in C_0^\gamma([0,T];E)$. Then the classical solution $u(\cdot)$ of the problem (5.5.2) lies in $C_0^\gamma([0,T];E)$ and the operator $u(t) = u(t; f, u^0)$ is continuous from $C_0^\gamma([0,T];E) \times D(A)$ to $C_0^\gamma([0,T];E)$.*

Theorem 5.11 (see [193, 200]). *Let $0 < \gamma$, $\alpha < 1$, $A \in \mathcal{A}^\alpha$, $f(\cdot) \in C_0^\gamma([0,T];E)$, and $u^0 \in D(A)$. Assume that $u(\cdot)$ is a unique classical solution of the problem (5.5.2). Then $Au(\cdot)$, $D^\alpha u(\cdot) \in C_0^\gamma([0,T];E)$, and*

$$\left\| D^\alpha u(\cdot) \right\|_{C_0^\gamma([0,T];E)} + \left\| Au(\cdot) \right\|_{C_0^\gamma([0,T];E)} \le C\left(\left\| f(\cdot) \right\|_{C_0^\gamma([0,T];E)} + \left\| Au^0 \right\|_E \right).$$

Remark 5.7. To prove the maximal regularity of the fractional Cauchy problem (5.5.2), one must consider the space $C_0^\gamma([0,T];E)$ rather than $C^\gamma([0,T];E)$. In [182], Li and Li studied the Hölder regularity for (5.5.2) in the space $C^\gamma([0,T];E)$. Under the additional condition $u^0 = f(0) = 0$, they proved that both functions $Au(\cdot)$ and $D^\alpha u(\cdot)$ belong to $C^\gamma([0,T];E)$. An excellent presentation of the maximal regularity in the space $C^\gamma([0,T];E)$ is given in [119].

5.5.2 Difference schemes for fractional equations

Now we consider an approximation of the fractional initial-value problem (5.5.2). The general discretization scheme for the problem (5.5.2) in a Banach space E_n is

$$\begin{cases} D^\alpha u_n(t) = A_n u_n(t) + f_n(t), & 0 \le t \le T, \\ u_n(0) = u_n^0. \end{cases} \tag{5.5.4}$$

The discretization of (5.5.4) can be performed by various difference schemes. First, we approximate the problem (5.5.4) by the implicit scheme

$$\Delta_{t_k}^\alpha \overline{U}_n(\cdot) = A_n \overline{U}_n(t_k) + f_n(t_k), \quad \overline{U}_n(0) = x_n. \tag{5.5.5}$$

where for grid function $\Theta_n(\cdot)$ we define finite difference derivative as

$$\Delta_{t_k}^\alpha \Theta_n(\cdot) = \frac{1}{\Gamma(2-\alpha)} \sum_{j=0}^{k-1} \left(t_{j+1}^{1-\alpha} - t_j^{1-\alpha} \right) \frac{\Theta_n(t_{k-j}) - \Theta_n(t_{k-j-1})}{\tau_n}.$$

It is well known that (5.5.5) can be split as follows:

$$\Delta_{t_k}^\alpha \overline{W}_n(\cdot) = A_n \overline{W}_n(t_k), \qquad\qquad \overline{W}_n(0) = x_n,$$
$$\Delta_{t_k}^\alpha \overline{V}_n(\cdot) = A_n \overline{V}_n(t_k) + f_n(t_k), \quad \overline{V}_n(0) = 0;$$

thus, $\overline{U}_n(t_k) = \overline{W}_n(t_k) + \overline{V}_n(t_k)$. Then from [197, Proposition 3.1 and Theorem 3.2] we obtain the following theorem.

Theorem 5.12 (see [200]). *For the implicit difference scheme* (5.5.5), *i.e., for the system*

$$\frac{1}{\Gamma(2-\alpha)} \sum_{j=0}^{k-1} b_j \frac{\overline{U}_n((k-j)\tau_n) - \overline{U}_n((k-j-1)\tau_n)}{\tau_n^\alpha}$$

$$= A_n \overline{U}_n(k\tau_n) + f_n(k\tau_n), \quad \overline{U}_n(0) = x_n, \qquad (5.5.6)$$

we have

$$\overline{U}_n(k\tau_n) = \sum_{j=1}^{k} c_j^{(k)} R^j x_n + \Gamma(2-\alpha)\tau_n^\alpha \sum_{j=1}^{k} \sum_{i=1}^{k-j+1} d_{i,j}^{(k)} R^i f_n(j\tau_n) \qquad (5.5.7)$$

for any $k \in \mathbb{N}$, where $c_j^{(k)}$ and $d_{i,j}^{(k)}$ are the same as in [197,198]. Moreover, for every $k \in \mathbb{N}$, $c_j^{(k)} > 0$, $j = 1, 2, \ldots, k$, $d_{i,j}^{(k)} \geq 0$, $i = 1, \ldots, k-j+1$, $j = 1, \ldots, k$, and

$$\sum_{j=1}^{k} c_j^{(k)} = 1, \quad \sum_{j=1}^{k} \sum_{i=1}^{k-j+1} d_{i,j}^{(k)} b_{j-1} = 1.$$

It is known that a bounded perturbation excludes the fractional well-posed Cauchy problem from the class of well-posed problems. At the same time, the principle of subordination allows us to conclude that the generator of C_0-semigroup A will generate a well-posed Cauchy problem for a fractional equation with any $0 < \alpha < 1$. Therefore, we will assume that the operator A generates a C_0-semigroup and the approximating operators satisfy the following condition (see [120, Theorem 3.1]):

(B) There are some constants $M \geq 1$ and ω independent of n such that $\|\exp(tA_n)\| \leq Me^{\omega t}$ for $t \geq 0$ and any $n \in \mathbb{N}$.

Theorem 5.13 (see [200]). *Assume that the C_0-semigroups e^{tA_n} satisfy Condition* (B) *and $\omega = 0$. Then the implicit difference scheme* (5.5.5) *is stable, i.e.,*

$$\sup_{1 \leq j \leq k} \left\|\overline{U}_n(j\tau_n)\right\| \leq M\|x_n\| + M\Gamma(1-\alpha)(k\tau_n)^\alpha \sup_{1 \leq j \leq k} \left\|f_n(j\tau_n)\right\|.$$

The next theorem gives us the order of convergence of the implicit difference scheme (5.5.5). We set $\overline{z}_{u_n}(k\tau_n) = u_n(k\tau_n) - \overline{U}_n(k\tau_n)$.

Theorem 5.14 (see [200]). *We have the following representation:*

$$\overline{z}_{u_n}(k\tau_n) = \Gamma(2-\alpha)\tau_n^\alpha \sum_{j=1}^{k} \sum_{i=1}^{k-j+1} d_{i,j}^{(k)} R^i \overline{r}_{u_n}(j\tau_n), \qquad (5.5.8)$$

where $d_{i,j}^{(k)}$ is the same as in [197,198]. Moreover, if the C_0-semigroups e^{tA_n} satisfy Condition (B) *and $\omega = 0$, then*

$$\sup_{1 \le j \le k} \left\| \bar{z}_{u_n}(j\tau_n) \right\| \le C(\alpha)\tau_n^\alpha m(x_n, f_n), \quad 0 \le k\tau_n \le T,$$

where

$$m(x_n, f_n) = \max \left\{ \|x_n\|, \|A_n x_n\|, \|A_n^2 x_n\|, \|f_n(\cdot)\|_{C^2([0,T];E_n)} \right\}.$$

Next, the explicit scheme

$$\Delta_{t_k}^\alpha U_n(\cdot) = A_n U_n(t_{k-1}) + f_n(t_{k-1}), \quad U_n(0) = x_n, \tag{5.5.9}$$

can be partitioned as follows:

$$\Delta_{t_k}^\alpha W_n(\cdot) = A_n W_n(t_k), \qquad\qquad W_n(0) = x_n,$$
$$\Delta_{t_k}^\alpha V_n(\cdot) = A_n V_n(t_k) + f_n(t_k), \quad V_n(0) = 0,$$

so that $U_n(t_k) = W_n(t_k) + V_n(t_k)$. Then we obtain the following theorem.

Theorem 5.15 (see [200]). *For the explicit scheme* (5.5.9), *i.e., for the scheme*

$$\frac{1}{\Gamma(2-\alpha)} \sum_{j=0}^{k-1} b_j \frac{U_n\big((k-j)\tau_n\big) - U_n\big((k-j-1)\tau_n\big)}{\tau_n^\alpha}$$
$$= A_n U_n\big((k-1)\tau_n\big) + f_n\big((k-1)\tau_n\big), \quad U_n(0) = x_n,$$

we have

$$U_n(k\tau_n) = \sum_{j=0}^{k} \bar{c}_j^{(k)} \bar{R}^j x_n + \Gamma(2-\alpha)\tau_n^\alpha \sum_{j=0}^{k-1} \sum_{i=0}^{k-j-1} \bar{d}_{i,j+1}^{(k+1)} \bar{R}^i f_n(j\tau_n)$$

for any $k \in \mathbb{N}$, where $\bar{c}_j^{(k)}$ and $\bar{d}_{i,j}^{(k)}$ are the same as in [197,198]. Moreover, for every $k \in \mathbb{N}$, we have

$$\bar{c}_j^{(k)} > 0, \qquad j = 0, 1, \dots, k,$$
$$\bar{d}_{i,j+1}^{(k+1)} \ge 0, \quad i = 0, \dots, k-j-1, \quad j = 0, \dots, k-1,$$

and

$$\sum_{j=0}^{k} \bar{c}_j^{(k)} = 1, \quad \sum_{j=0}^{k-1} \sum_{i=0}^{k-j-1} \bar{d}_{i,j+1}^{(k+1)} b_j = 1.$$

Remark 5.8. Here $V_n(k\tau_n)$ is somewhat different than in [197, 198] since $f_n(0) \ne 0$, whereas in [197, 198], $f_n(0) = 0$.

Theorem 5.16 (see [200]). *Let* $\alpha > 1/2$ *and let* e^{tA_n} *be the* C_0-*semigroups generated by the operators* A_n. *Assume that* e^{tA_n} *satisfy Condition* (B) *and* $\omega = 0$. *If*

$$\left\| \tau_n^{2\alpha-1} A_n^2 \right\| \le c,$$

where c *is independent of* n, *then for the scheme* (5.5.9) *we have*

$$\sup_{0 \le j \le k} \left\| U_n(j\tau_n) \right\| \le \bar{M} \|x_n\| + \bar{M}\Gamma(1-\alpha)(k\tau_n)^\alpha \sup_{0 \le j \le k-1} \left\| f_n(j\tau_n) \right\|.$$

Theorem 5.17 (see [200]). *For the explicit difference scheme* (5.5.9), *the difference* $z_{u_n}(k\tau_n) = u_n(k\tau_n) - U_n(k\tau_n)$ *can be represented as follows:*

$$z_{u_n}(k\tau_n) = \Gamma(2-\alpha)\tau_n^\alpha \sum_{j=1}^{k} \sum_{i=0}^{k-j} \bar{d}_{i,j}^{(k+1)} \bar{R}^i r_{u_n}(j\tau_n), \tag{5.5.10}$$

where

$$r_{u_n}(k\tau_n) = \Delta_{t_k}^\alpha u_n(\cdot) - \left(\mathbf{D}_t^\alpha u_n\right)(t_{k-1}),$$

$\bar{d}_{i,j}^{(k)}$ *are the same as in* [197, 198]. *If* $\alpha > 1/2$, *assume that the* C_0-*semigroups* e^{tA_n} *satisfy Condition* (B) *and* $\omega = 0$. *Moreover, if*

$$\left\| \tau_n^{2\alpha-1} A_n^2 \right\| \le c,$$

where c *is independent of* n, *then*

$$\sup_{1 \le j \le k} \left\| z_{u_n}(j\tau_n) \right\| \le C(\alpha)\tau_n^\alpha m(x_n, f_n).$$

Theorem 5.18 (see [200]). *We have the following representation for* $z_{u_n}(k\tau_n) = u_n(k\tau_n) - U_n(k\tau_n)$:

$$z_{u_n}(k\tau_n) = \Gamma(2-\alpha)\tau_n^\alpha \sum_{j=1}^{k} \sum_{i=0}^{k-j} \bar{d}_{i,j}^{(k+1)} \bar{R}^i r_{u_n}(j\tau_n), \tag{5.5.11}$$

where

$$r_{u_n}(k\tau_n) = \Delta_{t_k}^\alpha u_n(\cdot) - \left(\mathbf{D}_t^\alpha u_n\right)(t_{k-1}),$$

$\bar{d}_{i,j}^{(k)}$ *are the same as in* [197, 198]. *Assume that Condition* (B$_1$) *is fulfilled and* $\omega = 0$. *Assume also that*

$$\left\| \frac{\Gamma(2-\alpha)}{(1-b_1)}\tau_n^\alpha A_n \right\| \le c,$$

where $c < 1/(M+2)$ *is independent of* n. *Then, with* $m(x_n, f_n)$ *as in Theorem 5.14,*

$$\sup_{1 \le j \le k} \left\| z_{u_n}(j\tau_n) \right\| \le C(\alpha)\tau_n^\alpha m(x_n, f_n).$$

In [25] the relationship between fractional powers of positive operators and fractional derivatives of functions was discussed. It was proved that

$$\Lambda^\alpha \vartheta(\xi) = D^\alpha \vartheta(\xi), \qquad (5.5.12)$$

where $\Lambda \vartheta(\xi) = \vartheta'(\xi)$ with $D(\Lambda) = \{\vartheta(\xi) : \ \vartheta'(\xi) \in C[0,T], \ \vartheta(0) = 0\}$, Λ^α is the fractional-power operator of Λ, and $D_{t_k}^\alpha \vartheta(\cdot)$ is the Riemann–Liouville fractional derivative of function $\vartheta(\cdot) \in D(\Lambda)$. Here the operator Λ is the positive operator, since its spectrum $\sigma(\Lambda)$ if exists lies in the interior of the sector of aperture φ, $0 < \varphi < \pi$, symmetric with respect to the real axis, and since on the edges of this sector,

$$\Sigma_1 = \{\rho\exp(i\varphi) : 0 \le \rho < \infty\}, \quad \Sigma_2 = \{\rho\exp(-i\varphi) : 0 \le \rho < \infty\},$$

and, for λ outside of Σ_1,

$$\|(\lambda I - \Lambda)^{-1}\| \le C \frac{\varphi}{1 + |\lambda|}.$$

For such an operator (as in the case (1.1.3)) the negative fractional powers are defined by the formula

$$\Lambda^{-\alpha} = \frac{1}{2\pi i} \int_\Gamma \lambda^{-\alpha}(\lambda I - \Lambda)^{-1} d\lambda, \quad \Gamma = \Sigma_1 \cup \Sigma_2,$$

and positive fractional powers Λ^α is defined as the operators inverse to the negative powers. Regarding the question, which functions are fractionally differentiable see [290].

Based on Eq. (5.5.12), one can define the following Riemann–Liouville fractional difference derivative of order α, as the fractional power of order α of discretization of operators Λ by finite difference operator

$$\partial u_n(t_j) = \frac{u_n(t_j) - u(t_j - \tau_n)}{\tau_n}.$$

One can see that $\partial^\alpha \equiv \tilde{\Delta}_{t_k}^\alpha$ (see [25, 247]), where

$$\tilde{\Delta}_{t_k}^\alpha u_n(\cdot) = \frac{1}{\Gamma(1-\alpha)} \sum_{j=1}^k \frac{\Gamma(k-j-\alpha+1)}{(k-j)!} \cdot \frac{u_n(t_j) - u_n(t_{j-1})}{\tau_n^\alpha}. \qquad (5.5.13)$$

Therefore, we can approximate problem (5.5.4) by the following implicit difference scheme

$$\begin{cases} \tilde{\Delta}_{t_k}^\alpha \tilde{U}_n(\cdot) = A_n \tilde{U}_n(t_k) + f_n(t_k), \\ \tilde{U}_n(0) = u_n^0, \end{cases} \qquad (5.5.14)$$

with $t_k = k\tau_n \in [0,T]$, $k = 1, 2, \ldots, K$. Introduce the notation

$$b_j^{(k)} = \frac{\Gamma(k-j-\alpha+1)}{(k-j)! \cdot \Gamma(1-\alpha)}.$$

Then, in term of (5.5.14), we have

$$\sum_{j=1}^{k} b_j^{(k)} \frac{\widetilde{U}_n(t_j) - \widetilde{U}_n(t_{j-1})}{\tau_n^\alpha} = A_n \widetilde{U}_n(t_k) + f_n(t_k). \tag{5.5.15}$$

Setting $R = (I_n - \tau_n^\alpha A_n)^{-1}$, we have

$$\widetilde{U}_n(t_k) = Rb_1^{(k)}\widetilde{U}_n(0) + R\sum_{j=2}^{k}\left(b_j^{(k)} - b_{j-1}^{(k)}\right)\widetilde{U}_n(t_{j-1}) + R\tau_n^\alpha f_n(t_k). \tag{5.5.16}$$

Remark 5.9. Note that $b_j^{(k)}$ and $\tilde{\Delta}_{t_k}^\alpha \widetilde{U}_n(\cdot)$ here are different from b_j and $\Delta_{t_k}^\alpha \overline{U}_n(\cdot)$ in [197], where $b_j = (j+1)^{1-\alpha} - j^{1-\alpha}$. Thus, the implicit difference scheme (5.5.14) is different from the scheme (5.5.6).

Moreover, as recently learned from a private communication by Ru Liu, the scheme generated by (5.5.13) is equivalent to the scheme [37, (2.35)]. In both cases, i.e., in [25, 37], they just introduced the scheme, but did not prove its stability and did not get the order of convergence.

Theorem 5.19 (see [193]). *For the implicit difference scheme* (5.5.14), *we obtain the relation*

$$\widetilde{U}_n(k\tau_n) = \sum_{i=1}^{k} c_i^{(k)} R^i u_n^0 + \tau_n^\alpha \sum_{j=1}^{k}\sum_{i=1}^{k-j+1} d_{i,j}^{(k)} R^i f_n(j\tau_n),$$

where $c_1^{(k)} = b_1^{(k)}$, $d_{1,j}^{(k)} = 0$, $d_{1,k}^{(k)} = 1$,

$$c_i^{(k)} = \sum_{j=i}^{k}\left(b_j^{(k)} - b_{j-1}^{(k)}\right)c_{i-1}^{(j-1)}, \qquad i = 2, 3, \ldots, k,$$

$$d_{i,j}^{(k)} = \sum_{m=i+j-1}^{k}\left(b_m^{(k)} - b_{m-1}^{(k)}\right)d_{i-1,j}^{(m-1)}, \quad i = 2, 3, \ldots, k - j + 1,$$
$$j = 1, 2, \ldots, k - 1,$$

Moreover,

$$c_i^{(k)} > 0, \quad i = 1, 2, \ldots, k, \qquad \sum_{i=1}^{k} c_i^{(k)} = 1,$$

$$d_{i,j}^{(k)} \geq 0, \quad i = 1, 2, \ldots, k - j + 1, \quad j = 1, 2, \ldots, k,$$

$$\sum_{j=1}^{k}\sum_{i=1}^{k-j+1} d_{i,j}^{(k)} = \frac{(1+\alpha)(2+\alpha)\cdots(k-1+\alpha)}{(k-1)!}, \quad k \geq 2.$$

Theorem 5.20 (see [193]). *Assume that Condition* (B) *holds with* $\omega = 0$. *Then the implicit difference scheme* (5.5.14) *is stable, i.e.,*

$$\sup_{1 \le j \le k} \|\widetilde{U}_n(k\tau_n)\| \le M_1 \|u_n^0\| + M_1 \exp(\alpha)(k\tau_n)^\alpha \sup_{1 \le j \le k} \|f_n(j\tau_n)\|,$$

where $k\tau_n \in [0, T]$.

For the implicit difference scheme (5.5.15), which approximates the problem (5.5.4), we have the following assertion.

Theorem 5.21 (see [194]). *Let* $u_n(\cdot)$ *be the solution of problem* (5.5.4) *and* $\widetilde{U}_n^k$ *be the solution of implicit difference scheme* (5.5.14). *Assume that Condition* (B) *holds with* $\omega = 0$. *Also assume that* $u_n^0 \in D(A^\ell)$ *with* $\alpha\ell > 1$, $f_n(\cdot) \in C^1([0, T]; E_n)$ *hold. Then, there exists a positive constant* C *such that*

$$\|\widetilde{U}_n^k - u_n(t_k)\| \le C\tau_n^\alpha, \tag{5.5.17}$$

where the constant C *is independent of* n *and* $t_k \in [0, T]$.

Now we consider the following explicit difference scheme

$$\begin{cases} \widetilde{\Delta}_{t_n}^\alpha \widehat{U}_n(\cdot) = A_n \widehat{U}_n(t_{k-1}) + f_n(t_{k-1}), \\ \widehat{U}_n(0) = u_n^0. \end{cases} \tag{5.5.18}$$

The convergence of this explicit scheme (5.5.18) was also considered in [194].

Theorem 5.22. *Let* $u_n(\cdot)$ *be a solution of the problem* (5.5.4) *and* $\widehat{U}_n(\cdot)$ *be a solution of the explicit difference scheme* (5.5.18). *Assume that Condition* (B$_1$) *holds with* $\omega = 0$. *Also assume that*

$$\|\tau_n^\alpha A_n\| \le \frac{\alpha}{M+2}$$

with some $M \ge 1$ *and* $u_n^0 \in D(A^\ell)$ *with* $\alpha\ell > 1$, *and* $f_n(\cdot) \in C^1([0, T]; E_n)$. *Then there exists a positive constant* C *such that*

$$\|u_n(t_k) - \widehat{U}_n(t_k)\| \le C\tau_n^\alpha,$$

where C *is independent of* n *and* τ_n.

As we have seen, the order of convergence on the uniform grid is $O(\tau^\alpha)$. In the case of a nonuniform grid, the convergence rate can be improved (see [157]).

Chapter 6

Stable and Unstable Manifolds
for Fractional Semilinear Equations

This chapter is devoted to the numerical analysis of the abstract semilinear fractional problem

$$\mathcal{D}^\alpha u(t) = Au(t) + f(u(t)), u(0) = u^0,$$

in a Banach space E. We present a general approach to semidiscrete approximations of stable and unstable manifolds. The main assumption of our results are naturally satisfied, in particular, for operators with compact resolvents.

The behavior of solutions of semilinear problems in neighborhoods of hyperbolic equilibrium points of classical differential equations is a well-studied subject (see, e.g., [56, 66, 69, 110, 129, 223, 235, 277, 299, 310]) with various applications (see, e.g., [91, 155, 178, 179]). The situation changes significantly if instead of classical differential equations, we consider fractional differential equations.

For example, some sufficient conditions that provide the finite-time attractivity of solutions to semilinear fractional differential equations with nonlinearities of superlinear growth were established in [277].

One of the fundamental distinctions is the fact that solutions of fractional problems converge to equilibria only with polynomial rate instead of exponential rate like in the classical case. There are several papers on the problem of the existence of stable manifolds for semilinear fractional equations (see, e.g., [75, 77, 86, 259]). In [259], K. Sayevand and K. Pichaghchi tried to approximate local stable manifolds near hyperbolic equilibrium points of fractional differential equations and discussed the convergence of numerical schemes. None of the papers mentioned considered approximations of stable and unstable manifolds within general approximation schemes.

Despite the fact that the theory of nonlinear fractional differential equations is in its infancy, we mention the papers $[89, 90, 111, 180]$.

6.1   Setting of the Fractional Semilinear Problems

In this section, we study the semidiscrete approximation of the Cauchy problem

$$(\mathcal{D}^\alpha u)(t) = Au(t) + f(u(t)), \quad 0 \leq t \leq \infty, \quad u(0) = u^0, \tag{6.1.1}$$

where $\mathcal{D}^\alpha$, $0 < \alpha \leq 1$, is the Caputo–Dzhrbashyan derivative, the operator A generates an analytic and compact C_0-semigroup $\exp(\cdot A)$, and the function $f(\cdot)$ is sufficiently smooth. More precisely, we assume that $A : D(A) \subseteq E \to E$ is a closed linear operator such that

$$\left\|(\lambda I - A)^{-1}\right\|_{B(E)} \leq \frac{M}{1 + |\lambda|} \quad \text{for any } \lambda, \text{ Re } \lambda \geq 0, \tag{6.1.2}$$

and the function $f : E^\beta \subseteq E \to E$, $0 \leq \beta < 1$, is assumed bounded and continuously Fréchet differentiable.

Since the C_0-semigroup $\exp(\cdot A)$ exists, the resolution family $S_\alpha(t, A) \equiv E_\alpha(t^\alpha A)$, $t \geq 0$, is analytic in t (see [37, Theorem 3.3]) and we have (see [92, 148])

$$E_\alpha(t^\alpha A)x = \int_0^\infty t^{-\alpha}\Phi_\alpha(\xi t^{-\alpha})\exp(\xi A)x d\xi \quad \text{for any } x \in E \text{ and } t > 0, \tag{6.1.3}$$

where $\Phi_\alpha(\cdot)$ is a Wright-type function (see [37]):

$$\Phi_\alpha(0) = \frac{1}{\Gamma(1 - \alpha)}, \quad \Phi_\alpha(t) \geq 0, \quad t > 0, \quad \int_0^\infty \Phi_\alpha(t)\, dt = 1.$$

A solution of the problem (6.1.1) in the nonhomogeneous case is given by the formula

$$u(t) = E_\alpha(t^\alpha A)u^0 + \int_0^t P_\alpha(t - s, A)f(u(s))ds, \quad t \geq 0, \tag{6.1.4}$$

where

$$P_\alpha(t, A)x = \alpha \int_0^\infty \frac{s}{t^{\alpha+1}}\Phi_\alpha(st^{-\alpha})\exp(sA)x\, ds$$

$$= \alpha \int_0^\infty \frac{\tau}{t^{1-\alpha}}\Phi_\alpha(\tau)\exp(\tau t^\alpha A)x\, d\tau \tag{6.1.5}$$

for any $x \in E$, and $P_\alpha(t, A)$ is analytic in t for any $t > 0$ (see [148]). Moreover, (5.1.4) and (5.3.2) hold for any $\text{Re } \lambda > \omega$, $x \in E$. Note that in the scalar case, i.e., if $A = \gamma I$, we have $S_\alpha(t, \gamma) = E_\alpha(\gamma t^\alpha)$, where $E_\alpha(\cdot)$ is the Mittag-Leffler function

$$E_{\alpha,\beta}(z) = \sum_{n=0}^{\infty} \frac{z^n}{\Gamma(\alpha n + \beta)}, \quad E_\alpha(z) = E_{\alpha,1}(z).$$

Definition 6.1. A function $u(\cdot) \in C([0, \infty); E)$ is called a *mild solution* of the problem (6.1.1) if it satisfies Eq. (6.1.4).

Equilibrium solutions of (6.1.1) are solutions of the equation

$$Au + f(u) = 0. \tag{6.1.6}$$

We set

$$\Lambda_\alpha^+ = \left\{ \lambda \in \mathbb{C} : \ |\arg(\lambda)| < \frac{\alpha\pi}{2} \right\}.$$

In the case where the resolvent $(\lambda I - A)^{-1}$ is compact, the part σ^+ of the spectrum of the operator $A + f'(u^*)$, which is located strictly in the set $\sigma^+ \subset \Lambda_\alpha^+$, and it consists of a finite number of eigenvalues with finite multiplicity.

Definition 6.2. We say that a solution u^* of Eq. (6.1.6) is *hyperbolic point* if the spectrum $\sigma(A + f'(u^*))$ is disjoint from the boundary of Λ_α^+, i.e.,

$$\sigma(A + f'(u^*)) \cap \partial\Lambda_\alpha^+ = \varnothing.$$

6.2 Existence of Stable Manifolds

In this section, we also assume that Condition (F1) holds for a hyperbolic equilibrium point of (6.1.1).

Performing the change of variables $v(\cdot) = u(\cdot) - u^*$ in the problem (6.1.1), where u^* is a hyperbolic equilibrium, we obtain the problem

$$(\mathcal{D}^\alpha v)(t) = (A + f'(u^*))v(t) + f(v(t) + u^*) - f(u^*) - f'(u^*)v(t),$$
$$v(0) = u^0 - u^* = v^0,$$

We rewrite this problem in the form

$$(\mathcal{D}^\alpha v)(t) = A_{u^*}v(t) + F_{u^*}(v(t)), \quad v(0) = v^0, \quad t \geq 0, \tag{6.2.1}$$

where $A_{u^*} = A + f'(u^*)$ and $F_{u^*}(v(t)) = f(v(t) + u^*) - f(u^*) - f'(u^*)v(t)$. We note that

$$F_{u^*}(v(t)) = f(v(t) + u^*) - f(u^*) - f'(u^*)v(t) = o(\|v(t)\|_{E^\beta})$$

for small $\|v^0\|_{E^\beta}$. Since $f'(u^*) \in B(E^\beta, E)$, $0 \le \beta < 1$, the operator $A_{u^*} = A + f'(u^*)$ is the generator of an analytic C_0-semigroup (see [235]).

Let $U(\sigma^+) \subset \{\lambda \in \mathbb{C} : \text{Re}\,\lambda > 0\}$ be an open, connected neighborhood of σ^+ whose boundary $\partial U(\sigma^+) \subset \Lambda_\alpha^+$ is a closed rectifiable curve. We decompose E^β using the Riesz projection

$$P(\sigma^+) = P(\sigma^+, A_{u^*}) = \frac{1}{2\pi i} \int_{\partial U(\sigma^+)} (\zeta I - A_{u^*})^{-1}\, d\zeta \qquad (6.2.2)$$

determined by σ^+. Due to this definition and the analyticity of the C_0-semigroup $e^{tA_{u^*}}$, $t \in \mathbb{R}^+$, there are positive constants M and ω such that the inequalities (3.1.10) hold with $\theta = \beta$. Without loss of generality, we can adapt the norm in E^β such that

$$\|v\|_{E^\beta} = \max\left(\|P(\sigma+)v\|_{E^\beta}, \; \|(I - P(\sigma+))v\|_{E^\beta} \right).$$

It is clear that this norm is equivalent to the original norm in E^β.

We will need the following lemmas (see [151, 236]).

Lemma 6.1. Let $0 < \alpha < 1$ and $\omega > 0$. Then the following statements hold:

(i) $\quad E_\alpha(\omega t^\alpha) = \dfrac{1}{\alpha} e^{\omega^{1/\alpha}t} + O\left(\dfrac{1}{t^\alpha}\right)$ $\qquad\qquad$ as $t \to \infty$;

(ii) $\quad E_\alpha(\omega t^\alpha) \to 1$ $\qquad\qquad\qquad\qquad\qquad\qquad$ as $t \to 0$;

(iii) $\quad t^{\alpha-1} E_{\alpha,\alpha}(\omega t^\alpha) = \dfrac{1}{\alpha} \omega^{1/\alpha-1} e^{\omega^{1/\alpha}t} + O\left(\dfrac{1}{t^{\alpha+1}}\right)$ $\qquad$ as $t \to \infty$;

(iv) $\quad E_{\alpha,\alpha}(\omega t^\alpha) \to \dfrac{1}{\Gamma(\alpha)}$ $\qquad\qquad\qquad\qquad\qquad$ as $t \to 0$.

Lemma 6.2. Let $0 < \alpha < 1$ and $\omega > 0$. Then the following statements hold:

(i) $\quad E_\alpha(-\omega t^\alpha) = O\left(\dfrac{1}{t^\alpha}\right)$ $\qquad\qquad$ as $t \to \infty$;

(ii) $\quad E_\alpha(-\omega t^\alpha) \to 1$ $\qquad\qquad\qquad$ as $t \to 0$;

(iii) $\quad t^{\alpha-1} E_{\alpha,\alpha}(-\omega t^\alpha) = O\left(\dfrac{1}{t^{\alpha+1}}\right)$ $\qquad$ as $t \to \infty$;

(iv) $\quad E_{\alpha,\alpha}(-\omega t^\alpha) \to \dfrac{1}{\Gamma(\alpha)}$ $\qquad\qquad$ as $t \to 0$.

Lemma 6.3. Let $\exp(tA)$, $t \ge 0$, be an analytic C_0-semigroup such that

$$\|\exp(tA)\| \le M e^{-\omega t}$$

for $\omega > 0$, $t \geq 0$, and $0 < \alpha, \beta < 1$. Then

$$\left\|(-A)^\beta P_\alpha(t, A)x\right\| \leq \frac{C}{t^{1-\alpha(1-\beta)}} \|x\| \quad \text{for any } x \in E, \ t \to 0, \tag{6.2.3}$$

$$\left\|(-A)^\beta P_\alpha(t, A)x\right\| \leq \frac{C_1}{t^{1+\alpha}} \|x\| \qquad \text{for any } x \in E, \ t \to \infty. \tag{6.2.4}$$

Proof. From (6.1.5) we obtain

$$\left\|(-A)^\beta P_\alpha(t, A)x\right\| \leq \alpha \int_0^\infty \frac{s}{t^{\alpha+1}} \Phi_\alpha(st^{-\alpha}) \frac{M}{s^\beta} e^{-\omega s} \|x\| \, ds$$

$$\leq M \int_0^\infty \frac{(\xi t^\alpha)^{1-\beta}}{t^{\alpha+1}} \Phi_\alpha(\xi) e^{-\omega \xi t^\alpha} t^\alpha \, d\xi \, \|x\|$$

$$\leq C \int_0^\infty \frac{\xi^{1-\beta}}{t^{1-\alpha(1-\beta)}} \Phi_\alpha(\xi) e^{-\omega \xi t^\alpha} \, d\xi \, \|x\|.$$

Since

$$\left| e^{-\omega \xi t^\alpha} \right| \leq 1, \qquad \int_0^\infty \xi^{1-\beta} \Phi_\alpha(\xi) d\xi = \frac{\Gamma(2-\beta)}{\Gamma(1+\alpha(1-\beta))},$$

we obtain the estimate (6.2.3). To get (6.2.4), we again use (6.1.5) and write

$$\int_0^\infty \frac{(\xi t^\alpha)^{1-\beta}}{t^{\alpha+1}} \Phi_\alpha(\xi) e^{-\omega \xi t^\alpha} t^\alpha \, d\xi \, \|x\|$$

$$\leq C \int_0^\infty \frac{(\omega \xi t^\alpha)^{1-\beta}}{t^{1+\alpha}} \Phi_\alpha(\xi) e^{-\omega \xi t^\alpha} \, d(\omega \xi t^\alpha) \, \|x\| \leq \frac{C_1}{t^{1+\alpha}} \|x\|. \qquad \square$$

Now we split the problem (6.2.1) as follows:

$$w(t) = P(\sigma^+)v(t), \quad z(t) = (I - P(\sigma^+))v(t).$$

Hence, if $A_{u^*}^+$ is the restriction of the operator $(A + f'(u^*))$ to $P(\sigma^+)E^\beta$ and $A_{u^*}^-$ is the restriction of the operator $(A + f'(u^*))$ to $(I - P(\sigma^+))E^\beta$, we have

$$\left(\mathcal{D}^\alpha w\right)(t) = A_{u^*}^+ w(t) + H\big(w(t), z(t)\big), \tag{6.2.5}$$

$$\left(\mathcal{D}^\alpha z\right)(t) = A_{u^*}^- z(t) + G\big(w(t), z(t)\big), \tag{6.2.6}$$

where

$$H(w, z) = P(\sigma^+)f\big(w(t) + z(t) + u^*\big) - P(\sigma^+)f(u^*)$$
$$- P(\sigma^+)f'(u^*)\big(w(t) + z(t)\big),$$

$$G(w, z) = \big(I - P(\sigma^+)\big)f\big(w(t) + z(t) + u^*\big)$$
$$- \big(I - P(\sigma^+)\big)f(u^*) - \big(I - P(\sigma^+)\big)f'(u^*)\big(w(t) + z(t)\big).$$

Hence we see that the functions H and G and their derivatives vanish at $(0,0)$.

Similarly to Theorem 3.1 we find that a mild solution of $(6.2.5)$–$(6.2.6)$, which tends to zero as $t \to \infty$, is a vector-function $v(t) = \begin{pmatrix} w(t) \\ z(t) \end{pmatrix}$ satisfying the formal equation (see [229, formula (3.7)])

$$v(t) = \Theta(\eta, v(\cdot))(t), \tag{6.2.7}$$

but looks different when compared with Theorem 3.1. We consider $(6.2.7)$ in the space

$$\begin{aligned}
\Upsilon &= C_\gamma\Big([0,\infty), P(\sigma^+)E^\beta \times \big(I - P(\sigma^+)\big)E^\beta\Big) \\
&= \Big\{ v(\cdot) \in C\Big([0,\infty), P(\sigma^+)E^\beta \times \big(I - P(\sigma^+)\big)E^\beta\Big) : \\
&\qquad v(\cdot) \text{ is bounded on } [0,\infty)\Big\}
\end{aligned}$$

with the supremum norm

$$\|v\|_{C_\gamma} = \|v\|_\Upsilon = \sup_{t \geq 0} \frac{\|v(t)\|_{E^\beta}}{E_\alpha(-\gamma t^\alpha)}$$

for some $\gamma > 0$.

Then we show that $\Theta(\eta, \cdot) : \Upsilon \to \Upsilon$. To do this, we take $q(\cdot) \in \Upsilon$, so $q(\cdot)$ is a function decreasing to zero as $O(t^{-\alpha})$, $t \to \infty$. Clearly, $\Theta(\eta, q(\cdot))$ is a continuous function in t. We show that $\lim_{t \to \infty} \Theta(\eta, q(\cdot))(t) = 0$ and the order of convergence as $t \to \infty$ is $-\alpha$. Any projector P_j is finite-dimensional, i.e., the generalized eigenspace corresponding to an eigenvalue λ_j is finite-dimensional of dimension $\operatorname{rank} P_j = n_j$ and, therefore, for $j = 1, \dots, K$, the jth equation involves the operator $A_{u^*} P_j$, which either has the form $\lambda_j I_{n_j \times n_j}$ in the case where λ_j is a simple eigenvalue or the form $\lambda_j I_{n_j \times n_j} + N_{n_j \times n_j}$, where $N_{n_j \times n_j} \in \mathbb{R}^{n_j \times n_j}$ is a nilpotent matrix satisfying the condition $(N_{n_j \times n_j})^{n_j} = 0$. We consider these cases separately.

First, we consider the case of a simple eigenvalue λ_j, i.e., $A_{u^*} P_j = \lambda_j I_{n_j \times n_j}$. For any given $\varepsilon > 0$ and $q(\cdot)$, there exists $t_\varepsilon \in (1,\infty)$ such that $\varepsilon \geq 1/t_\varepsilon^\alpha$ and

$$\big\| G(q(s)) \big\|_{(I - P(\sigma^+))E^\beta} = O(s^{-\alpha}) \quad \text{for all } s \geq t_\varepsilon.$$

As was shown in [37, Corollary 3.2] and [189], one has

$$\Big\| E_\alpha(t^\alpha A_{u^*}^-) \Big\| \leq M E_\alpha(-\omega t^\alpha) = O(t^{-\alpha}) \quad \text{as } t \to \infty,$$

so it follows that

$$\lim_{t \to \infty} E_\alpha(t^\alpha A_{u^*}^-)\eta = 0$$

with order $-\alpha$ as $t \to \infty$ for any $\eta \in (I - P(\sigma^+))E^\beta$. For $t \geq t_\varepsilon$, from Lemma 6.3 we have

$$\left\| (-A_{u^*})^\beta \int_0^t P_\alpha(t - s, A_{u^*}^-) G(q(s)) \, ds \right\|$$

$$\leq C \left\| \int_0^{t_\varepsilon} (-A_{u^*})^\beta P_\alpha(s, A_{u^*}^-) G\big(q(t - s)\big) \, ds \right.$$

$$\left. + \int_{t_\varepsilon}^t (-A_{u^*})^\beta P_\alpha(s, A_{u^*}^-) G\big(q(t - s)\big) \, ds \right\|$$

$$\leq C \int_0^1 \frac{1}{s^{1-\alpha(1-\beta)}} \big\| G(q(t - s)) \big\| \, ds$$

$$+ C \int_1^{t_\varepsilon} \frac{1}{s^{1+\alpha}} \big\| G(q(t - s)) \big\| \, ds$$

$$+ C \int_{t_\varepsilon}^t \frac{1}{s^{1+\alpha}} \big\| G(q(t - s)) \big\| \, ds \leq C \left(\varepsilon + \frac{1}{t^\alpha} \right).$$

Hence for given $\varepsilon > 0$,

$$\left\| \int_0^t (-A_{u^*})^\beta P_\alpha(t - s, A_{u^*}^-) \, G(q(s)) ds \right\| \leq C\varepsilon + O(t^{-\alpha}) \quad \text{as } t \to \infty.$$

It is clear from Lemma 6.1(i) that

$$E_\alpha(t^\alpha \lambda_j) \lambda_j^{1/\alpha - 1} \int_t^\infty \exp\left(-\lambda_j^{1/\alpha} \tau \right) \big\| P_j H\big(w(\tau), z(\tau) \big) \big\| \, d\tau = O\left(\frac{1}{t^\alpha} \right)$$

as $t \to \infty$ since $\big\| P_j H(w(\tau), z(\tau)) \big\| \leq C\tau^{-\alpha}$.

To estimate the first term, for any $\varepsilon > 0$ we find $t_\varepsilon > 0$ such that $\big\| P_j H(w(\tau), z(\tau)) \big\| \leq \varepsilon$ for $\tau > t_\varepsilon$.

We obtain that the Fréchet derivative $\Theta_q'(\eta_0, q_0) : \Upsilon \to \Upsilon$ is given by

$$(\Theta_q'(\eta_0, q_0)h)(t) = 0$$

for any $h = \begin{pmatrix} h_1(t) \\ h_2(t) \end{pmatrix} \in \Upsilon$, since the functions G and H have derivatives equal to 0 at zero. Therefore, the operator $(I - \Theta_q'(\eta_0, q_0))^{-1}$ exists and is continuous. From [168, Theorem 54.2] it follows that Eq. (6.2.7) has a unique solution.

To examine $A_{u^*}P_j = \lambda_j I_{n_j \times n_j} + N_{n_j \times n_j}$ in the case where λ_j is not simple, following [77], we apply for a given small $\delta > 0$ a linear, invertible transformation, which transforms $A_{u^*}P_j$ into $\lambda_j I_{n_j \times n_j} + \delta N_{n_j \times n_j}$.

Under appropriate assumptions, for some $\epsilon > 0$, the part of the spectrum of the operator A_{u^*} given by $\{\lambda \in \sigma(A_{u^*}) \setminus \Lambda_\alpha^+,\ \mathrm{Re}\,\lambda \geq -\epsilon\}$, consists of a finite number of eigenvalues with finite multiplicity. We denote it by Σ_α^+. Since the resolvent of the operator A is compact, it can happen that such finite set (perhaps, empty) Σ_α^+ exists.

As was shown in [76] for any $\lambda_0 \in \Sigma_\alpha^+$ the solution of (6.1.1) on a finite dimensional subspace $P(\lambda_0)E^\beta$ converge to zero as $t \to \infty$, i.e., $\begin{pmatrix} w(t) \\ z(t) \end{pmatrix} \to 0$ as $t \to \infty$.

We gave a sketch of the proof of the following theorem.

Theorem 6.1 (see [229]). *Let u^* be a hyperbolic equilibrium for (6.1.1). Assume that the operator A generates an analytic C_0-semigroup $\exp(\cdot A)$ and Condition* (F1) *is fulfilled. Then for some sufficiently small $\delta > 0$, there exists a Lipschitz continuous function*

$$\pounds : \left\{ \eta \in \left(I - P(\sigma^+)\right)E^\beta :\ \|\eta\|_{E^\beta} \leq \delta \right\} \to P(\sigma^+)E^\beta, \quad \pounds(\eta) = w(0),$$

such that the local stable manifold $W_\delta^{st}(u^*)$ is given by

$$W_\delta^{st}(u^*) = \left\{ \left(\eta, \pounds(\eta)\right) + u^* :\ \eta \in \left(I - P(\sigma^+)\right)E^\beta,\ \|\eta\|_{E^\beta} \leq \delta \right\}.$$

6.3   Approximation of Stable Manifolds

Here we describe the discretization of semilinear fractional equations on a general approximation scheme and discuss the approximation of stable manifolds.

Theorem 6.2 (see [14, 96, 192, 196, 203, 307]). *Let the operator A be the generator of an analytic C_0-semigroup $\exp(tA)$ in a Banach space E such that $\|\exp(tA)\| \leq Me^{-\omega t}$, $t \geq 0$. Then the compactness of $R(\lambda^\alpha; A)$ for some $\lambda^\alpha \in \rho(A)$ is equivalent to the compactness of $S_\alpha(t, A)$ for any $t > 0$. Moreover, the compactness of $R(\lambda^\alpha; A)$ for some $\lambda^\alpha \in \rho(A)$ is equivalent to the compactness of $P_\alpha(t, A)$ for any $t > 0$.*

Proof. We prove only the last statement. We know that the function $P_\alpha(\cdot, A)$ is analytic and compact. Then, due to the representation

$$R(\lambda^\alpha; A) = \int_0^\infty e^{-\lambda t} P_\alpha(t, A)\, dt,$$

we conclude that the operator function

$$R_{q,Q}(\lambda) = \int_q^Q e^{-\lambda t} P_\alpha(t, A)\, dt$$

approximates the resolvent:

$$\left\| R(\lambda^\alpha; A) - R_{q,Q}(\lambda) \right\| \to 0 \quad \text{as } q \to 0 \text{ and } Q \to \infty.$$

The operator-function $R_{q,Q}(\lambda)$ is compact (see [300]); hence $R(\lambda^\alpha; A)$ is compact as the uniform limit of compact operators. Conversely, we can write (see [37])

$$P_\alpha(t, A) = \frac{1}{2\pi i} \int_\Gamma e^{\lambda t} R(\lambda^\alpha; A)\, d\lambda \quad \text{for any } t > 0,$$

where Γ is a positively oriented path, which is the boundary of $\omega + \Sigma_{\beta + \pi/2}$. Since $R(\lambda^\alpha; A)$ is compact for any $\lambda^\alpha \in \rho(A)$, it follows that $P_\alpha(t, A)$ is compact for any $t > 0$. $\square$

Theorem 6.3 (see [196]). *Assume that the operators A_n and A generate analytic C_0-semigroups. Assume that Conditions* (A) *and* (B_1) *hold with $\omega_2 < 0$ and the resolvents $R(\lambda^\alpha; A_n)$ and $R(\lambda^\alpha; A)$ are compact. Then the following conditions are equivalent:*

(i) $R(\lambda^\alpha; A_n) \xrightarrow{\;PP\;} R(\lambda^\alpha; A)$ *compactly for some* $\lambda^\alpha \in \rho(A) \cap \bigcap_n \rho(A_n)$;

(ii) $S_\alpha(t, A_n) \xrightarrow{\;PP\;} S_\alpha(t, A)$ *compactly for any* $t > 0$;

(iii) $P_\alpha(t, A_n) \xrightarrow{\;PP\;} P_\alpha(t, A)$ *compactly for any* $t > 0$.

Proof. The equivalence of (i)$\Leftrightarrow$(ii) was proved in [196]. To prove the implication (i)$\Leftarrow$(iii), we need to show that the measure of noncompactness $\mu\big(\{R(\lambda^\alpha; A_n)x_n\}\big)$ vanishes for any $\{x_n\}$ such that $\|x_n\|_{E_n} = O(1)$. We have

$$\mu\big(\{R(\lambda^\alpha; A_n)x_n\}\big) = \mu\left(\left\{\int_0^\infty e^{-\lambda t} P_\alpha(t, A_n)x_n dt\right\}\right)$$

$$\leq \mu\left(\left\{\int_0^q e^{-\lambda t} P_\alpha(t, A_n)x_n dt\right\}\right)$$

$$+ \mu\left(\left\{\int_Q^\infty e^{-\lambda t} P_\alpha(t, A_n)x_n dt\right\}\right)$$

$$+ \mu\left(\left\{\int_q^Q e^{-\lambda t} P_\alpha(t, A_n)x_n dt\right\}\right),$$

where q is small and Q is sufficiently large so that the first and second terms are less than any small $\varepsilon > 0$. We know that $\{P_\alpha(t, A_n)\}$ is equicontinuous on $[q, Q]$. Then the third term is zero. The implication (i)$\Rightarrow$(iii) follows from the compact convergence $\exp(tA_n) \xrightarrow{\mathcal{PP}} \exp(tA)$, $t > 0$, and the representation (6.1.5). $\qquad\square$

As in (4.2.4), we define

$$p_n^\beta = (-A_n)^{-\beta} p_n(-A)^\beta \in B(E^\beta, E_n^\beta).$$

Assume that the operators A_n and A satisfy Condition (6.1.2) and Conditions (A) and (B$_1$).

In the Banach spaces E_n^β, $0 \le \beta < 1$, consider the family of fractional problems

$$\begin{aligned}
\left(\mathcal{D}^\alpha u_n\right)(t) &= A_n u_n(t) + f_n\left(u_n(t)\right), \quad t \ge 0, \\
u_n(0) &= u_n^0 \in E_n^\beta,
\end{aligned} \tag{6.3.1}$$

where $u_n^0 \xrightarrow{\mathcal{P}^\beta} u^0$, the operators (A_n, A) are compatible, and the functions $f_n(\cdot) : E_n^\beta \to E_n$ are globally bounded and globally Lipschitz continuous uniformly in $n \in \mathbb{N}$ and continuously Fréchet differentiable. We also assume that the operators A_n generate analytic C_0-semigroups and

$$f_n(x_n) \xrightarrow{\mathcal{P}} f(x), \quad f_n'(x_n) \xrightarrow{\mathcal{P}^\beta \mathcal{P}} f'(x) \quad \text{as } x_n \xrightarrow{\mathcal{P}^\beta} x.$$

Under the above assumptions, a mild solution $u_n(\cdot)$ of (6.3.1) is defined for all $t \ge 0$ by the equation

$$u_n(t) = S_\alpha(t, A_n) u_n^0 + \int_0^t P_\alpha(t - s, A_n) f_n(u_n(s)) ds, \quad t \ge 0. \tag{6.3.2}$$

In the case where $\Delta_{cc} \ne \varnothing$, the convergence $A_n^{-1} f_n(\cdot) \xrightarrow{\mathcal{PP}} A^{-1} f(\cdot)$ is compact. For $n \ge n_0$, it follows from the results of [288] that

$$1 = \gamma\left(I + A^{-1} f'(u^*); \partial\Omega\right) = \gamma\left(I_n + A_n^{-1} f_n(u_n^*); \partial\Omega_n\right)$$

for any $u^* \in E$ with $n \ge n_0$ and then the equations $u_n^* = -A_n^{-1} f_n(u_n^*)$ have solutions such that $u_n^* \xrightarrow{\mathcal{P}} u^*$. Clearly, $u_n^* \xrightarrow{\mathcal{P}^\beta} u^*$.

Now we consider a hyperbolic equilibrium u^* of (6.1.1) and hyperbolic equilibria $u_n^* \xrightarrow{\mathcal{P}^\beta} u^*$ of (6.3.1); moreover, in the case where $\theta = \beta$ and u_n^* are hyperbolic points of (6.3.1) we assume that Condition (F3) is fulfilled.

Remark 6.1. We get

$$\left\| F_{u_n^*, n}'(q_n) \right\| \le c_\rho \|q_n\|_{E_n^\beta},$$

where $c_\rho \to 0$ as $\rho \to 0$. Here $\rho > 0$ is the radius of balls $\mathcal{U}_{E_n^\beta}(0; \rho)$ for which there exists $\delta > 0$ such that

$$\sup_{n \in \mathbb{N}} \sup_{\|q_n\|_{E_n^\beta} \le \delta} \left\| f_n'\left(q_n + p_n^\beta u^*\right) - f_n'\left(p_n^\beta u^*\right) \right\|_{B(E_n^\beta, E_n)} \le \rho.$$

Now we consider the family of fractional problems (6.3.1) in a neighborhood of a hyperbolic equilibrium point u_n^*. In this case we have

$$\left(\mathcal{D}^\alpha v_n\right)(t) = A_{u_n^*, n} v_n(t) + F_{u_n^*, n}\left(v_n(t)\right), \quad t \ge 0, \tag{6.3.3}$$
$$v_n(0) = v_n^0,$$

where

$$A_{u_n^*, n} = A_n + f_n'(u_n^*),$$
$$F_{u_n^*, n}\left(v_n(t)\right) = f_n\left(v_n(t) + u_n^*\right) - f_n(u_n^*) - f_n'(u_n^*) v_n(t).$$

Similarly to (4.2.12), we decompose E_n^β using the projection operators $P_n(\sigma_n^+)$ and similarly to the operator $\Theta(\eta, u(\cdot))$ (see (6.2.7)), we define the operator $\Theta_n(\eta_n, u_n(\cdot))$. Normally, we assume that $\eta_n \xrightarrow{\mathcal{P}^\beta} \eta$ or even $\eta_n = p_n^\beta \eta$. Rewriting (6.3.1) for $v_n(t) = u_n(t) - u_n^*$, we can deal with a neighborhood of u_n^*; then we rewrite (6.3.3) as follows:

$$\begin{cases} D^\alpha w_n(t) = \left(A_n + f_n'(u_n^*)\right) P_n(\sigma_n^+) w_n(t) + H_n\left(w_n(t), z_n(t)\right), \\ D^\alpha z_n(t) = \left(A_n + f_n'(u_n^*)\right)\left(I_n - P_n(\sigma_n^+)\right) z_n(t) + G_n\left(w_n(t), z_n(t)\right), \end{cases} \tag{6.3.4}$$

where $w_n(\cdot) \in P_n(\sigma_n^+) E_n^\beta$ and $z_n(\cdot) \in \left(I_n - P_n(\sigma_n^+)\right) E_n^\beta$. We write

$$H_n(w_n, z_n) = P_n(\sigma_n^+)$$
$$\times \left(f_n(w_n + z_n + u_n^*) - f_n(u_n^*) - f_n'(u_n^*)(w_n + z_n) \right),$$
$$G_n(w_n, z_n) = \left(I_n - P_n(\sigma_n^+)\right)$$
$$\times \left(f_n(w_n + z_n + u_n^*) - f_n(u_n^*) - f_n'(u_n^*)(w_n + z_n) \right).$$

Hence we see that $H_n(0,0) = 0$ and $G_n(0,0) = 0$. From the continuous differentiability of H_n and G_n and the estimates

$$\begin{cases} \left\| e^{t A_{u_n^*, n}} z_n \right\|_{E_n^\beta} \le M_2 e^{-\gamma t} \|z_n\|_{E_n^\beta}, & t \ge 0, \\ \left\| e^{t A_{u_n^*, n}} z_n \right\|_{E_n^\beta} \le M_2 t^{-\beta} e^{-\gamma t} \|z_n\|_{E_n}, & t > 0, \\ \left\| e^{t A_{u_n^*, n}} w_n \right\|_{E_n^\beta} \le M_2 e^{\gamma_1 t} \|w_n\|_{E_n^\beta}, & t \le 0, \end{cases} \tag{6.3.5}$$

which hold uniformly in n under the analyticity conditions of the C_0-semigroup $e^{t A_{u_n^*, n}}$ and the condition $\Delta_{cc} \ne \varnothing$, we obtain that for given $\rho > 0$, there exist $n_0 > 0$ and $\delta > 0$ such that if

$$\|w_n\|_{P_n(\sigma_n^+) E_n^\beta} + \|z_n\|_{(I_n - P_n(\sigma_n^+)) E_n^\beta} < \delta$$

and $n \geq n_0$, we have

$$\left\| H_n(w_n, z_n) \right\|_{P_n(\sigma_n^+)E_n} \leq \rho, \quad \left\| G_n(w_n, z_n) \right\|_{E_n} \leq \rho,$$

$$\left\| H_n(w_n, z_n) - H_n(\tilde{v}_n, \tilde{z}_n) \right\|_{P_n(\sigma_n^+)E_n}$$

$$\leq \rho \left(\left\| w_n - \tilde{w}_n \right\|_{P_n(\sigma_n^+)E_n^\beta} + \left\| z_n - \tilde{z}_n \right\|_{(I_n - P_n(\sigma_n^+))E_n^\beta} \right), \tag{6.3.6}$$

$$\left\| G_n(w_n, z_n) - G_n(\tilde{w}_n, \tilde{z}_n) \right\|_{E_n}$$

$$\leq \rho \left(\left\| w_n - \tilde{w}_n \right\|_{P_n(\sigma_n^+)E_n^\beta} + \left\| z_n - \tilde{z}_n \right\|_{(I_n - P_n(\sigma_n^+))E_n^\beta} \right).$$

The fact that we can choose ρ and δ uniformly for $n \geq n_0$ satisfying the above inequalities, is the key point to obtain that the local stable manifolds are defined in a small "neighborhood" of the equilibrium point u_n^* uniformly for $n \geq n_0$.

Conditions (B$_1$) and $\Delta_{cc} \neq \varnothing$ imply that $P_n(\sigma_n^+) \xrightarrow{\mathcal{PP}} P(\sigma^+)$ compactly; therefore, as was shown in [56,69], the spectra $\sigma(A_{u_n^*,n}^+)$ approximate the spectrum $\sigma(A_{u_*}^+)$ and $\dim P_n(\sigma_n^+) = \dim P(\sigma^+)$ for $n \geq n_0$.

Introduce the notation

$$\Omega^\alpha = \left\{ v(\cdot) \in C([0,\infty); E^\beta) : \ \|v(t)\|_{E^\beta} = O(t^{-\alpha}) \text{ as } t \to \infty \right\},$$

where the norm of $v(\cdot) \in \Omega^\alpha$ is given by

$$\|v\|_{\Omega^\alpha} = \sup_{t \in [0,\infty)} \frac{\|v(t)\|_{E^\beta}}{E_\alpha(-\gamma t^\alpha)} < \infty$$

for some $\gamma > 0$.

Theorem 6.4 (see [229]). *Assume that the operator A generates an analytic and compact C_0-semigroup. Assume that Conditions (F1) and (6.1.2) are fulfilled. Then the operator*

$$K : \Omega^\alpha \to \Omega^\alpha,$$

$$(Ku)(t) \equiv S_\alpha(t, A)u^0 + \int_0^t P_\alpha(t - s, A)f(u(s))\, ds, \quad t \in [0,\infty), \tag{6.3.7}$$

is compact.

Proof. Let $\{u_k(\cdot)\}$ be a set of functions $u_k(\cdot) \in \Omega^\alpha$ such that

$$\|u_k(\cdot)\|_{\Omega^\alpha} \leq \text{const}, \quad k \in \mathbb{N}.$$

From Condition (F1) we know that $f(\cdot)$ is bounded, and then we obtain that the set of functions $\{(Ku_k)(\cdot)\}$ is uniformly bounded, where $(Ku_k)(\cdot) \in \Omega^\alpha$. Let T be such that

$$\left\| f(u_k(t)) \right\| \leq \varepsilon, \quad t \geq T.$$

For $0 < t_1 < t_2 \leq T$, we have

$$\left\|(Ku_k)(t_2) - (Ku_k)(t_1)\right\|_{E^\beta}$$

$$= \left\|(-A)^\beta \left(S_\alpha(t_2, A)u_k^0 - S_\alpha(t_1, A)u_k^0 \right.\right.$$

$$\left.\left. + \int_0^{t_2} P_\alpha(t_2 - s, A)f\big(u_k(s)\big)\, ds - \int_0^{t_1} P_\alpha(t_1 - s, A)f\big(u_k(s)\big)\, ds \right)\right\|$$

$$\leq \left\|\big(S_\alpha(t_2, A) - S_\alpha(t_1, A)\big)(-A)^\beta u_k^0\right\|$$

$$+ \int_0^{t_1 - \delta} \left\|(-A)^\beta \big(P_\alpha(t_2 - s, A) - P_\alpha(t_1 - s, A)\big)\right\| \cdot \left\|f(u_k(s))\right\|\, ds$$

$$+ \int_{t_1 - \delta}^{t_1} \left\|(-A)^\beta \big(P_\alpha(t_2 - s, A) - P_\alpha(t_1 - s, A)\big)\right\| \cdot \left\|f(s, u_k(s))\right\|\, ds$$

$$+ \int_{t_1}^{t_2} \left\|(-A)^\beta P_\alpha(t_2 - s, A)\right\| \cdot \left\|f(s, u_k(s))\right\|\, ds$$

$$\leq c\left\|S_\alpha(t_2, A) - S_\alpha(t_1, A)\right\| + \bar{M}(t_1 - \delta)$$

$$\times \sup_{s \in [0, t_1 - \delta]} \left\|(-A)^\beta \big(P_\alpha(t_2 - s, A) - P_\alpha(t_1 - s, A)\big)\right\|$$

$$+ 2M\bar{M}\delta + M\bar{M}(t_2 - t_1)^{\alpha(1-\beta)} \to 0 \quad \text{as } |t_2 - t_1| \to 0.$$

Lemma 6.3 implies that

$$\left\|\int_T^t P_\alpha(t - s, A)f(u(s))\, ds\right\| \leq C\varepsilon \quad \text{as } t \geq T.$$

This means that $\{(Ku_k)(\cdot)\}$ is equicontinuous. Now we prove that the sequence $\{(Ku_k)(t)\}$ is relatively compact for any $t > 0$. From Theorem 6.2 we conclude that $S_\alpha(t, A)$ and $P_\alpha(t, A)$ are compact for any $t > 0$. The results of [300] imply that

$$\int_0^T P_\alpha(t - s, A)f(u(s))\, ds$$

is compact for any finite T and

$$\left\|\int_T^t P_\alpha(t - s, A)f(u(s))\, ds\right\|$$

is sufficiently small for large T. This means that

$$\left\{\int_0^t (-A)^\beta P_\alpha(t - s, A)f(u_k(s))ds\right\} \tag{6.3.8}$$

is relatively compact for any $t > 0$. Obviously, the set (6.3.8) for $t = 0$ is also relatively compact. This implies that for any t, $\{(Ku_k)(t)\}$ is a relatively compact set in E^β. Then it follows from the generalized Arzelà–Ascoli theorem that, for each $0 \le t \le \infty$, the sequence $\{(Ku_k)(t)\}$ is relatively compact in E^β and, therefore, $\{(Ku_k)(\cdot)\}$ is a relatively compact sequence in Ω^α. Hence, K is a compact operator. $\qquad\square$

We also define the operators

$$K_n : \Omega_n^\alpha \to \Omega_n^\alpha,$$

$$(K_n u_n)(t) \equiv S_\alpha(t, A_n)u_n^0$$

$$+ \int_0^t P_\alpha(t - s, A_n)f_n(u_n(s))\,ds, \quad t \in [0, \infty), \tag{6.3.9}$$

where

$$\Omega_n^\alpha \equiv \left\{ u_n(\cdot) \in C\big([0, \infty); E_n^\beta\big) : \; \|u_n\|_{F_n} = \sup_{t \in [0, \infty)} \frac{\|u_n(t)\|_{E_n^\beta}}{E_\alpha(-\gamma t^\alpha)} < \infty \right\}$$

with some $\gamma > 0$.

Theorem 6.5. *Assume that Conditions* (F1)–(F3) *and* (B$_1$) *are satisfied and* $\Delta_{cc} \ne \varnothing$. *Then the convergence* $K_n \xrightarrow{\;\mathcal{P}^\beta \mathcal{P}^\beta\;} K$ *as* $u_n^0 \xrightarrow{\;\mathcal{P}^\beta\;} u^0$ *is compact.*

Proof. The operators K_n are compact. If $u_n(\cdot) \xrightarrow{\;\mathcal{P}^\beta\;} u(\cdot)$, then our assumption in (6.3.1) implies that $f_n(u_n(t)) \xrightarrow{\;\mathcal{P}\;} f(u(t))$ uniformly in $t \in [0, \infty)$. One can achieve

$$\sup_{t \in [0, T]} \left\| (K_n u_n)(t) - p_n^\beta (Ku)(t) \right\|_{E_n^\beta}$$

$$\le \sup_{t \in [0, \infty)} \left\| S_\alpha(t, A_n)u_n^0 - p_n^\beta S_\alpha(t, A)u^0 \right\|_{E_n^\beta}$$

$$+ \sup_{t \in [0, \infty)} \left\| \int_0^t P_\alpha(t - s, A_n)f_n\big(s, u_n(s)\big)\,ds \right.$$

$$\left. - p_n^\beta \int_0^t P_\alpha(t - s, A)f\big(u(s)\big)\,ds \right\|_{E_n^\beta} \to 0.$$

Hence we get $K_n \xrightarrow{\;\mathcal{P}^\beta \mathcal{P}^\beta\;} K$. Let $\{u_n(\cdot)\}$ be an arbitrary sequence of functions $u_n(\cdot) \in \Omega_n^\alpha$ such that $\|u_n(\cdot)\|_{\Omega_n^\alpha} = O(1)$. Next we show that

$$\mu\big(\{(K_n u_n)(t)\}\big) = 0 \quad \text{for all } t \in [0, \infty).$$

Indeed,

$$
\begin{aligned}
\mu(\{(-A_n)^\beta (K_n u_n)(t)\}) \\
\leq \mu(\{S_\alpha(t, A_n)(-A_n)^\beta u_n^0\}) \\
+ \mu\left(\left\{\int_0^t (-A_n)^\beta P_\alpha(t-s, A_n) f_n(u_n(s)) ds\right\}\right) \\
\leq \mu(\{S_\alpha(t, A_n)(-A_n)^\beta u_n^0\}) \\
+ \mu\left(\left\{\int_0^{t-\delta} (-A_n)^\beta P_\alpha(t-s, A_n) f_n(u_n(s)) ds\right\}\right) \\
+ \sup_{t \in [0,\infty)} \left\| \int_{t-\delta}^t P_\alpha(t-s, A_n) f_n(u_n(s)) ds \right\|_{E_n^\beta}.
\end{aligned}
$$

Obviously, the first term is zero. Since

$$
\left\| P_\alpha(t, A_n) \right\|_{B(E_n^\beta)} \leq O\left(\frac{1}{t^{1-\alpha(1-\beta)}}\right)
$$

by Lemma 6.3 and

$$
\sup_{t \in [0,T]} \left\| f_n(u_n(t)) \right\|_{E_n} \leq \bar{M},
$$

we see that the third term can be made less than ε if we choose δ sufficiently small. It follows from Condition (F3) that $\{f_n(u_n(s))\}$ is equicontinuous on $[0, t-\delta]$. We know that $\{(-A_n)^\beta P_\alpha(t-s, A_n)\}$ is also equicontinuous on $[0, t-\delta]$. Thus, we see that

$$
\mu\left(\left\{\int_0^{t-\delta} (-A_n)^\beta P_\alpha(t-s, A_n) f_n(u_n(s)) ds\right\}\right) = 0.
$$

Hence, $\{(K_n u_n)(\cdot)\}$ is P^β-compact. Thus, $K_n \xrightarrow{\; P^\beta P^\beta \;} K$ compactly. $\quad\square$

Now we consider the convergence of solutions of the equations

$$
(K_n u_n)(t) = u_n(t) \tag{6.3.10}
$$

to the solution of the equation

$$
(Ku)(t) = u(t). \tag{6.3.11}
$$

Theorem 6.6 (see [229]). *Let $\Delta_{cc} \neq \varnothing$. Assume that Conditions (A), (B$_1$), and (F1)–(F3) are fulfilled and $f_n(x_n) \xrightarrow{\; P \;} f(x)$ as $x_n \xrightarrow{\; P^\beta \;} x$. Then $u_n(t) \xrightarrow{\; P^\beta \;} u(t)$ uniformly with respect to $t \in [0, \infty)$ as $u_n^0 \xrightarrow{\; P^\beta \;} u^0$ for solutions of Eqs. (6.3.10) and (6.3.11), respectively.*

Proof. It follows from the assumption on $f(\cdot)$ that K has no fixed point on the boundary ∂Q_r (this follows from the uniqueness of the solution of Eq. (6.3.11)), where

$$Q_r = \left\{ u(\cdot): \sup_{t \in [0,\infty)} \|u(t) - u^*\|_{E_n^\beta} < r \right\}.$$

Since the families $S_\alpha(\cdot, A)$ and $P_\alpha(\cdot, A)$ are analytic and compact, the proof of the fact that $\gamma(I - K; \partial Q_r) = 1$ literally repeats the proof of [295, Theorem 14.2.3]. From [288] we conclude that

$$\gamma(I - K; \partial Q_r) = \gamma(I_n - K_n; \partial Q_{n,r}) \quad \text{for } n \geq n_0.$$

This means that there is a sequence $\{u_n(\cdot)\}$, $n \in \mathbb{N}$, of solutions of the problems (6.3.10), $u_n(\cdot) \in \Upsilon_n$, which is P^β-compact; hence one can write $u_n(t) \xrightarrow{\mathcal{P}^\beta} u(t)$ uniformly in $t \in [0, \infty)$ as $n \to \infty$. □

Theorem 6.7 (see [229]). *Let the conditions of Theorem 6.4 be fulfilled. Then the operator* $\Theta(\eta, \cdot) : \Upsilon \to \Upsilon$ *is compact.*

Proof. First, we mention that the projector $P(\sigma^+)$ is finite-dimensional. For any P_j, the function

$$\int_0^t P_\alpha(t - s, A_{u*}^+) P_j H\big(w(s), z(s)\big) ds$$

$$- E_\alpha(t^\alpha A_{u*}^+) \int_0^\infty Q\big(e^{-\lambda_1^{1/\alpha} \tau}, \tau\big) P_j H\big(w(\tau), z(\tau)\big) d\tau$$

is an equibounded and equicontinuous function in $t > 0$. Thus, by the Arzelà–Ascoli theorem, the part of the operator Θ corresponding to $P(\sigma^+)$ is compact. To prove the compactness of the part

$$E_\alpha\big(t^\alpha A_{u*}^-\big)\eta + \int_0^t P_\alpha(t - s, A_{u*}^-) G\big(w(s), z(s)\big) ds$$

of the operator Θ, we just need to quote Theorem 6.4. □

Here we formulate the statement on the approximation of stable manifolds in the case of fractional differential equations.

Theorem 6.8. *Let $\Delta_{cc} \neq \varnothing$ and Conditions (F1)–(F3) be satisfied. Then the convergence*

$$\Theta_n(\eta_n, \cdot) \to \Theta(\eta, \cdot) \quad \text{as } \eta_n \xrightarrow{\mathcal{P}^\beta} \eta$$

of the operators $\Theta_n(\eta_n, \cdot) : \Upsilon_n \to \Upsilon_n$ and $\Theta(\eta, \cdot) : \Upsilon \to \Upsilon$ is compact.

Proof. We recall that one can write

$$\left\|q_n(\cdot)\right\|_{\Upsilon_n} = \left\|\begin{pmatrix} q_n^1(\cdot) \\ q_n^2(\cdot) \end{pmatrix}\right\|_{\Upsilon_n}$$

$$= \sup_{t \geq 0} \left\|q_n^1(t)\right\|_{P_n(\sigma_n^+)E_n^\beta} + \sup_{t \geq 0} \left\|q_n^2(t)\right\|_{(I_n - P_n(\sigma_n^+))E_n^\beta}.$$

Thus, the discrete convergence $\Upsilon_n \ni q_n(\cdot) \to q(\cdot) \in \Upsilon$ means that

$$\sup_{0 \leq t < \infty} \left\|q_n^1(t) - p_n^\beta q^1(t)\right\|_{P_n(\sigma_n^+)E_n^\beta}$$

$$+ \sup_{0 \leq t < \infty} \left\|q_n^2(t) - p_n^\beta q^2(t)\right\|_{(I_n - P_n(\sigma_n^+))E_n^\beta} \to 0 \quad \text{as } n \to \infty.$$

Now we verify that the condition $\|q_n(\cdot)\|_{\Upsilon_n} = O(1)$ implies that the sequence $\{\Theta_n(\eta_n, q_n(\cdot))\}$ is P^β-compact. We apply the Arzelà–Ascoli theorem to the sequences

$$(-A_n)^\beta E_\alpha\big(t^\alpha A_{u_n^*}^-\big)\eta_n + (-A_n)^\beta \int_0^t P_\alpha\big(t - s, A_{u_n^*}^-\big)G_n\big(q_n^1(s), q_n^2(s)\big)\,ds$$

and

$$(-A_n)^\beta \left(\int_0^t P_\alpha(t - s, A_{u_n^*}^+)P_{n,j}\tilde{H}_n\big(q_n^1(s), q_n^2(s)\big)\,ds \right.$$

$$\left. - E_\alpha(t^\alpha A_{u_n^*}^+) \int_0^\infty Q\big(e^{-\lambda_1^{1/\alpha}s}, s\big)P_{n,j}\tilde{H}_n\big(q_n^1(s), q_n^2(s)\big)\,ds \right).$$

Due to (6.3.6) we have

$$\left\|(-A_n)^\beta \int_0^t P_\alpha(t - s, A_n)\tilde{H}_n\big(q_n^1(s), q_n^2(s)\big)\,ds\right\| \leq \varepsilon \quad \text{for } t \geq t_\varepsilon.$$

Then we apply the Arzelà–Ascoli theorem to the sets of functions

$$\left\{ (-A_n)^\beta \int_0^t P_\alpha\Big(t - s, A_n + f_n'(u_n^*)\Big)P_n(\sigma_n^+)\tilde{H}_n(q_n^1(s), q_n^2(s))ds \right\},$$

$$\left\{ (-A_n)^\beta \int_0^t P_\alpha\big(t - s, A_{u_n^*}^+\big)\big(I_n - P_n(\sigma_n^+)\big)G_n\big(q_n^1(s), q_n^2(s)\big)ds \right\}$$

on $[0, t_\varepsilon]$. This can be done as in Theorem 6.5. $\qquad\square$

Theorem 6.9 (see [229]). *Let $\Delta_{cc} \neq \varnothing$ and u^* be a hyperbolic equilibrium point of the problem* (6.1.1). *Then for sufficiently small $\delta > 0$, Eq.* (6.2.7) *has a solution $\zeta^*(\cdot)$ such that the equations*

$$\zeta_n(t) = \Theta_n\big(\eta_n, \zeta_n(t)\big)$$

have solutions $\zeta_n^*(\cdot)$ for $n \geq n_0$ and $\zeta_n^*(t) \to \zeta^*(t)$ uniformly in $t \in [0, \infty)$ as $n \to \infty$.

Proof. We know that $\left|\operatorname{ind}(\zeta_0, I - \Theta)\right| = 1$ for $\zeta_0 = 0$. According to [168, Theorem 54.1], there exist ρ and $\delta > 0$ such that $\left|\operatorname{ind}(\zeta, I - \Theta)\right| = 1$ for $\|\eta - 0\|_{E^\beta} \leq \delta$ and $\|\zeta - 0\|_\Upsilon \leq \rho$. By Theorem 6.8, the operators $\Theta_n \to \Theta$ converge compactly, so

$$\gamma(I_n - \Theta_n, \partial Q_n) = \operatorname{ind}(\zeta, I - \Theta)$$

and, therefore, the equation

$$\zeta_n(t) = \Theta_n(\eta_n, \zeta_n(t))$$

has at least one solution $\zeta_n^*(\cdot)$ in any set Q_n for $n \geq n_0$. This sequence $\{\zeta_n^*(\cdot)\}$ is discretely compact and $\zeta_n^*(t) \to \zeta^*(t)$ uniformly in $t \geq 0$ as $n \to \infty$. □

Proposition 6.1. *Assume that u^* is a hyperbolic equilibrium for the problem (6.1.1) and the spectrum of the operator $A + f'(u^*) : D(A) \subset E \to E$ does not contain zero. Assume that $\Delta_{cc} \neq \varnothing$ and $\left\|f_n'(u_n) - f_n'(p_n^\beta u^*)\right\| \to 0$ as $\left\|u_n - p_n^\beta u^*\right\|_{E_n^\beta} \to 0$. Then there exist $\delta > 0$ and $n_0 > 0$ such that u_n^* has a local stable manifold $W_{loc}^{st}(u_n^*) \subset E_n^\beta$ for any $n \geq n_0$, and if we denote its δ-part by*

$$W_{n,\delta}^{st}(u_n^*) = \left\{ v_n(\cdot) \in W_{loc}^{st}(u_n^*) : \left\|v_n(t) - u_n^*\right\|_{E_n^\beta} < \delta, \ t \geq 0 \right\}, \quad n \geq n_0,$$

then $W_{n,\delta}^{st}(u_n^*)$ converges to $W_\delta^{st}(u^*)$ as $n \to \infty$, that is, for any $t \in [0, \infty)$,

$$\sup_{v_n \in W_{n,\delta}^{st}(u_n^*)} \inf_{v \in W_\delta^{st}(u^*)} \left\|v_n(t) - p_n^\beta v(t)\right\|_{E_n^\beta}$$

$$+ \sup_{v \in W_\delta^{st}(u^*)} \inf_{v_n \in W_{n,\delta}^{st}(u_n^*)} \left\|v_n(t) - p_n^\beta v(t)\right\|_{E_n^\beta} \to 0$$

as $n \to \infty$.

Proof. We show that for a suitably small $\rho > 0$, there is a stable manifold for u_n^*

$$W_n^{st} = \left\{ (w_n(\cdot), z_n(\cdot)) : z_n(\cdot) \in (I_n - P_n(\sigma_n^+))E_n^\beta, \ w_n(\cdot) \in P_n(\sigma_n^+)E_n^\beta \right\}.$$

For the vector $\zeta_n(t) = \begin{pmatrix} w_n(t) \\ z_n(t) \end{pmatrix}$, we see from (6.3.6) the following. There exists positive δ such that the operators $\Theta_n(\eta_n, \cdot)$ are Fréchet differentiable in the balls $\mathcal{U}_{E_n^\beta}(p_n^\beta \zeta^*; \delta)$ and for any $\varepsilon > 0$, there is $\delta_\varepsilon > 0$ such that

$$\left\|\Theta_n'(\eta_n, \zeta_n) - \Theta_n'(\eta_n, p_n^\beta \zeta^*)\right\| \leq \varepsilon$$

whenever

$$\left\|\zeta_n(\cdot) - p_n^\beta \zeta^*(\cdot)\right\|_{\Upsilon_n} \leq \delta_\varepsilon.$$

It is also clear that as $n \to \infty$, the conditions

$$\left\|p_n^\beta \zeta^*(\cdot) - \Theta_n\left(\eta_n, p_n^\beta \zeta^*(\cdot)\right)\right\|_{\Upsilon_n} \to 0,$$

$$\left\|\left(I_n - \Theta_n'\left(\eta_n, p_n^\beta \zeta^*(\cdot)\right)\right)^{-1}\right\| \leq \text{const}$$

are satisfied. Then by [288, Theorem 2], there exist $n_0 \in \mathbb{N}$ and $0 < \delta_0 \leq \delta$ such that for $n \geq n_0$, Eqs. (6.3.4) provide a unique solution $\zeta_n^*(\cdot)$ in the balls $\mathcal{U}_{E_n^\beta}(p_n^\beta \zeta^*; \delta_0)$ and $\zeta_n^*(\cdot) \to \zeta^*(\cdot)$ as $n \to \infty$. $\qquad \square$

6.4 Existence of Unstable Manifolds

The part $\sigma^+ \subset \Lambda_\alpha^+$ of the spectrum of the operator $A + f'(u^*)$, which is located in Λ_α^+, is called the unstable spectrum. According to [76], the part of spectrum outside of Λ_α^+, i.e., the points of $\mathbb{C} \setminus \overline{\Lambda_\alpha^+}$, is called a stable spectrum of A_{u^*}. We say that u^* is a hyperbolic equilibrium point if there is no spectrum of A_{u^*} on the boundary of Λ_α^+. Let u^* be a hyperbolic equilibrium point and $U(\sigma^+) \subset \Lambda_\alpha^+$ be an open connected neighborhood of σ^+ which has a closed rectifiable curve $\partial U(\sigma^+)$ as a boundary. In the same way as in (6.2.2), we decompose E^β using the Riesz projection

$$P(\sigma^+) := P(\sigma^+, A_{u^*}) := \frac{1}{2\pi i} \int_{\partial U(\sigma^+)} (\zeta I - A_{u^*})^{-1} d\zeta \qquad (6.4.1)$$

associated to σ^+. We mention here that the set $U(\sigma^+)$ is different from the set in (3.1.5) (see also (4.2.12)) since $\sigma^+ \subset \Lambda_\alpha^+$. We can equip E^β with the norm $\|v\|_{E^\beta} = \max\{\|P(\sigma^+)v\|_{E^\beta}, \|(I - P(\sigma^+))v\|_{E^\beta}\}$, which is equivalent to the original norm on E^β.

Equation (4.1.6) can now be split into two parts: $w(t) = P(\sigma^+)v(t)$ and $z(t) = (I - P(\sigma^+))v(t)$. We denote by $A_{u^*}^+$ the restriction of $(A + f'(u^*))$ to $P(\sigma^+)E^\beta$ and we denote by $A_{u^*}^-$ the restriction of $(A + f'(u^*))$ to $(I - P(\sigma^+))E^\beta$. This yields

$$\begin{aligned}
(\mathbf{D}_t^\alpha w)(t) &= A_{u^*}^+ w(t) + H(w(t), z(t)), \\
(\mathbf{D}_t^\alpha z)(t) &= A_{u^*}^- z(t) + G(w(t), z(t)),
\end{aligned} \qquad (6.4.2)$$

where

$$\begin{aligned}
H(w, z) = \ & P(\sigma^+)f(w(t) + z(t) + u^*) \\
& - P(\sigma^+)f(u^*) - P(\sigma^+)f'(u^*)(w(t) + z(t)),
\end{aligned}$$

$$G(w, z) = (I - P(\sigma^+))f(w(t) + z(t) + u^*)$$
$$- (I - P(\sigma^+))f(u^*) - (I - P(\sigma^+))f'(u^*)(w(t) + z(t)).$$

Hence, we have that, at $(0,0)$ the functions H and G as well as their derivatives are zero at $(0,0)$.

The mild solution of (5.1.1) for the Caputo derivative with $0 < \alpha \leq 1$ and initial data $u(\tau) = u^\tau$ is given by the formula (see [109, (8.2.16)])

$$u(t) = E_\alpha((t - \tau)^\alpha A)u(\tau) + \int_\tau^t (t - s)^{\alpha-1} E_{\alpha,\alpha}((t - s)^\alpha A)f(u(s))ds.$$

One of the goals here is to show that for a suitably small $\rho > 0$, there exists a local unstable manifold in the ball $\|v\| \leq \rho$ for the hyperbolic equilibrium point u^*

$$W_\rho^u(u^*) = \Big\{w(t) + z(t) + u^*, \ w(0) = \eta \in P(\sigma^+)E^\beta :$$

$$z(t) \in (I - P(\sigma^+))E^\beta, \ w(t) \in P(\sigma^+)E^\beta, \ \|v(t)\| \leq \rho, \ t \in (-\infty, 0]\Big\}.$$

A solution $(w(t), z(t))$ of Eq. (6.4.2) in the unstable manifold must tend to zero as $t \to -\infty$. Since

$$z(t) = E_\alpha((t - \tau)^\alpha A_{u^*}^-)z(\tau)$$
$$+ \int_\tau^t (t - s)^{\alpha-1} E_{\alpha,\alpha}((t - s)^\alpha A_{u^*}^-)\, G(w(s), z(s))ds,$$

letting $\tau \to -\infty$, we have that

$$z(t) = \int_{-\infty}^t P_\alpha(t - s, A_{u^*}^-)\, G(w(s), z(s))ds, \quad -\infty < t \leq 0.$$

If $\operatorname{Re}\lambda \geq -\epsilon$ for $\lambda \in \sigma(A_{u^*}^-) \cap \Sigma_\alpha^+$ and some $\epsilon > 0$, then the function $E_\alpha((t - \tau)^\alpha\lambda)$ tends to zero as $\tau \to -\infty$ (see [109, (3.4.30)]). The convergence $E_\alpha((t - \tau)^\alpha A_{u^*}^-) \to 0$ as $\tau \to -\infty$ on $(I - P(\Sigma_\alpha^+) - P(\sigma+))E^\beta$ is obtained in [37, Corollary 3.2]. The definition of $P(\Sigma_\alpha^+)$ for Σ_α^+ is the same as the definition of $P(\sigma^+)$ for σ^+.

The function $w(\cdot)$ satisfies the equation

$$w(t) = E_\alpha((t - \tau)^\alpha A_{u^*}^+)w(\tau)$$
$$+ \int_\tau^t (t - s)^{\alpha-1} E_{\alpha,\alpha}((t - s)^\alpha A_{u^*}^+)\, H(w(s), z(s))ds,$$

and we know that $\sigma^+ \subset \Lambda_\alpha^+$. Then the operator $E_\alpha((t - \tau)^\alpha A_{u^*}^+)^{-1}$ is well-defined (see [156]) since the arguments of all roots of the Mittag-Leffler function lie the interval $(\alpha\pi/2, \alpha\pi)$ (see [240]). By the spectrum mapping theorem this means that zero does not appear in the spectrum of the matrix $E_\alpha(t^\alpha A_{u*}^+)$ if the spectrum of the matrix A_{u*}^+ belongs to the sector of complex numbers with an argument between $-\alpha\pi/2$ and $\alpha\pi/2$. So the inverse matrix $(E_\alpha(t^\alpha A_{u*}^+))^{-1}$ exists and, therefore,

$$E_\alpha\big((t-\tau)^\alpha A_{u*}^+\big)^{-1} w(t) = w(\tau) + E_\alpha\big((t-\tau)^\alpha A_{u*}^+\big)^{-1}$$
$$\times \int_\tau^t (t-s)^{\alpha-1} E_{\alpha,\alpha}\big((t-s)^\alpha A_{u*}^+\big) H(w(s),z(s))ds. \qquad (6.4.3)$$

From (6.4.3), setting $t = 0$ and changing τ for t, we have

$$w(t) = E_\alpha\big((-t)^\alpha A_{u*}^+\big)^{-1} w(0) - E_\alpha\big((-t)^\alpha A_{u*}^+\big)^{-1}$$
$$\times \int_t^0 (-s)^{\alpha-1} E_{\alpha,\alpha}\big((-s)^\alpha A_{u*}^+\big) H(w(s),z(s))ds. \qquad (6.4.4)$$

The mild solution of (6.4.2) is a vector-valued function

$$v(t) = \begin{pmatrix} w(t) \\ z(t) \end{pmatrix},$$

which satisfies the equation

$$v(t) = \Theta(\eta, v(t))$$

$$\equiv \begin{pmatrix} E_\alpha\big((-t)^\alpha A_{u*}^+\big)^{-1}\eta - E_\alpha\big((-t)^\alpha A_{u*}^+\big)^{-1} \\ \times \int_t^0 (-s)^{\alpha-1} E_{\alpha,\alpha}\big((-s)^\alpha A_{u*}^+\big) H(w(s),z(s))ds \\ \int_{-\infty}^t P_\alpha(t-s, A_{u*}^-) G(w(s),z(s))ds \end{pmatrix}, \qquad (6.4.5)$$

where $-\infty < t \le 0$ and $\eta = P(\sigma^+)w(0)$.

Now, let us consider (6.4.5) in the space of continuous functions

$$\Upsilon = C\big((-\infty,0]; P(\sigma^+)E^\beta\big) \times C\big((-\infty,0]; (I - P(\sigma^+))E^\beta\big)$$

with the sup norm and the condition $w(-\infty) = z(-\infty) = 0$. Let us show that $\Theta(\eta, \cdot) : \Upsilon \to \Upsilon$.

Take $\zeta(\cdot) \in \Upsilon$ then, according to (4.1.6) and (6.4.2), for any $\epsilon > 0$ there exists a number $-\infty < t_\epsilon < 0$ such that

$$\|G(\zeta(s))\|_{(I-P)E^\beta}, \ \|H(\zeta(s))\|_{PE^\beta} \le \epsilon, \quad s \le t_\epsilon.$$

According to Lemma 3.3 in [263] for the second coordinate of $\Theta(\eta, \cdot)$ the integral

$$(-A)^\beta \int_{-\infty}^t P_\alpha(t-s, A_{u*}^-) G(\zeta(s))ds \to 0$$

as $t \to -\infty$. To see this for the points $\lambda \in \sigma(A_{u^*}^-)$ of the spectrum with $\operatorname{Re}\lambda < 0$ we put $t - s = \xi$ and get

$$\left\| (-A)^\beta \int_0^{+\infty} P_\alpha(\xi, A_{u^*}^-) G(\zeta(t-\xi)) d\xi \right\|$$

$$\leq \left\| (-A)^\beta \int_0^{|t_\epsilon|} P_\alpha(\xi, A_{u^*}^-) G(\zeta(t-\xi)) d\xi \right\|$$

$$+ \left\| (-A)^\beta \int_{|t_\epsilon|}^{+\infty} P_\alpha(\xi, A_{u^*}^-) G(\zeta(t-\xi)) d\xi \right\|$$

$$\leq c\epsilon + c \left. \left(\frac{1}{\xi^\alpha} \right) \right|_{|t_\epsilon|}^{\infty} \to 0$$

as $t \to -\infty$, $\epsilon \to 0$, and $t = 2t_\epsilon$. Some points $\tilde{\lambda} \in \sigma(A_{u^*}^-)$ of the spectrum of the operator $A_{u^*}^-$ could have the property $\operatorname{Re}\tilde{\lambda} \geq 0$, but $\tilde{\lambda} \notin \Lambda_\alpha^+$. The number of such points is finite and they do not have accumulation points. So, for any such point $\tilde{\lambda} \in \Sigma_\alpha^+$ using the estimates of [76, Proposition 4(ii)], we obtain

$$\int_0^\infty \xi^{\alpha-1} E_{\alpha,\alpha}\left(\xi^\alpha \tilde{\lambda}\right) G(\zeta(t-\xi)) d\xi \to 0 \quad \text{as } t \to -\infty. \tag{6.4.6}$$

The operating with such points $\tilde{\lambda}$ to get (6.4.6) one has to follow the way which we describe in details when we will talk about points of σ^+.

Remark 6.2. The previous manipulation with $(-A)^\beta$ and the families generated, for example, by $A_{u^*}^-$ can be easily done owing to [56, Proposition 4.1 and Remark 4.2]. As was proved there, the operator $(-A)^\beta(-A_{u^*}^-)^{-\beta}$ is bounded. So one can write $(-A)^\beta = (-A)^\beta(-A_{u^*}^-)^{-\beta}(-A_{u^*}^-)^\beta$ for integrals, which are considered on the subspace $(I - P(\Sigma_\alpha^+) - P(\sigma+))E^\beta$.

The first coordinate of $\Theta(\eta, \cdot)$ is defined by (6.4.4) with finite-dimensional projector

$$P(\sigma^+) = \sum_{j=1}^K P_j.$$

Any projector P_j is finite-dimensional, i.e., the generalized eigenspace corresponding to the eigenvalue $\lambda_j \in \Lambda_\alpha^+$ is finite-dimensional of dimension rank $P_j = n_j$ and, therefore, for $j = 1, \ldots, K$ the jth equation involves the operator $A_{u^*} P_j$, which, after transformation $T^{-1} A_{u^*} P_j T$, takes either the form $\lambda_j I_{n_j \times n_j}$ in the case where λ_j is a semi-simple eigenvalue, or the form

$\lambda_j I_{n_j \times n_j} + N_{n_j \times n_j}$ with a nilpotent matrix $N_{n_j \times n_j} \in \mathbb{R}^{n_j \times n_j}$ satisfying the condition $(N_{n_j \times n_j})^{n_j} = 0$. We consider these cases separately.

In the case of $\lambda_j I_{n_j \times n_j}$, we have

$$E_\alpha\big((-t)^\alpha \lambda_j I_{n_j \times n_j}\big)^{-1} P_j \eta - E_\alpha\big((-t)^\alpha \lambda_j I_{n_j \times n_j}\big)^{-1}$$
$$\times \int_t^0 (-s)^{\alpha-1} E_{\alpha,\alpha}\big((-s)^\alpha \lambda_j I_{n_j \times n_j}\big) P_j H(\zeta(s)) ds \to 0 \qquad (6.4.7)$$

as $t \to -\infty$. Indeed, by [78, Lemma 2], the first term in (6.4.7) vanishes as $t \to -\infty$ for any bounded η. For the second term in (6.4.7), due to the asymptotic from [109, Example 3.9] for $\lambda_j \in \Lambda_\alpha^+$, we obtain $E_\alpha((-t)^\alpha \lambda_j)^{-1} \sim e^{\lambda_j^{1/\alpha} t} \to 0$ as $t \to -\infty$. From [78, Lemma 2], under the sign of integral we obtain

$$\frac{1}{\alpha}(\lambda_j)^{1/\alpha-1} e^{\lambda_j^{1/\alpha} s}, \quad s \le t_\epsilon.$$

So, for the integral term in (6.4.7), this leads to an integral of the type

$$\int_0^t e^{\lambda^{1/\alpha}(t-s)} H(\zeta(s)) ds.$$

Since $\|H(\zeta(s))\| \le \epsilon$ as $s \le t_\epsilon$, the proof is complete.

To consider the case of $A_{u^*} P_j = \lambda_j I_{n_j \times n_j} + N_{n_j \times n_j}$ in case λ_j is not semi-simple we follow [77], and we apply for a given small $\delta > 0$ a linear, invertible transformation $Q_j = \text{diag}(1, \delta, \ldots, \delta^{n_j-1})$ which transforms $A_{u^*} P_j$ into

$$Q_j^{-1}(\lambda_j I_{n_j \times n_j} + N_{n_j \times n_j}) Q_j = \lambda_j I_{n_j \times n_j} + \delta N_{n_j \times n_j}.$$

Applying this transformation to the differential equation, it leaves all but the jth equation unchanged, and transforms the term

$$\big[\lambda_j I_{n_j \times n_j} + N_{n_j \times n_j}\big] P_j(w, z) + P_j H(w, z)$$

in the jth equation into

$$\lambda_j I_{n_j \times n_j} P_j(w, z) + \big[\delta N_{n_j \times n_j} P_j(w, z) + (TQ_j)^{-1} P_j H(TQ_j \zeta(s))\big]. \quad (6.4.8)$$

The sum of the second and third term in brackets has a small Lipschitz constant for small $\delta > 0$ and the problem is reduced to the semi-simple case.

Now we formulate the main result.

Theorem 6.10 (see [230]). *Let u^* be a hyperbolic equilibrium for (5.1.1). Assume that the operator A generates an analytic, compact C_0-semigroup*

$\exp(\cdot A)$ *and Condition* (F1) *is satisfied. Then for some suitably small* $\delta > 0$, *there exists a continuous function*

$$\pounds : \left\{ \eta \in P(\sigma^+)E^\beta : \|\eta\|_{E^\beta} \leq \delta \right\} \to (I - P(\sigma^+))E^\beta, \quad \pounds(\eta) = z(0),$$

such that the local unstable manifold $W_\delta^u(u^*)$ *is given by*

$$W_\delta^u(u^*) = \left\{ (\eta, \pounds(\eta)) + u^* : \eta \in P(\sigma^+)E^\beta, \|\eta\|_{E^\beta} \leq \delta \right\}.$$

Proof. It has been proved already that $\Theta(\eta, \cdot) : \Upsilon \to \Upsilon$. We set

$$\eta = \eta_0 = 0, \quad v(t) = v_0(t) = \begin{pmatrix} w_0(t) \\ z_0(t) \end{pmatrix} = 0, \quad t \in (-\infty, 0];$$

then $v_0(t) = \Theta(0, v_0(\cdot))(t)$ and the operator $\Theta(\cdot, \cdot)$ is continuous in both arguments. Using the notation

$$\Theta'_{1,1}h_1 = -E_\alpha((-t)^\alpha A_{u^*}^+)^{-1}$$
$$\times \int_t^0 (-s)^{\alpha-1} E_{\alpha,\alpha}((-s)^\alpha A_{u^*}^+) H'_w(w(s), z(s))h_1(s)ds,$$

$$\Theta'_{1,2}h_2 = -E_\alpha((-t)^\alpha A_{u^*}^+)^{-1}$$
$$\times \int_t^0 (-s)^{\alpha-1} E_{\alpha,\alpha}((-s)^\alpha A_{u^*}^+) H'_z(w(s), z(s))h_2(s)ds,$$

$$\Theta'_{2,1}h_1 = \int_{-\infty}^t P_\alpha(t - s, A_{u^*}^-) G'_w(w(s), z(s))h_1(s)ds,$$

$$\Theta'_{2,2}h_2 = \int_{-\infty}^t P_\alpha(t - s, A_{u^*}^-) G'_z(w(s), z(s))h_2(s)ds,$$

we obtain that the Fréchet derivative $\Theta'_v(\eta_0, v_0) : \Upsilon \to \Upsilon$ is given by

$$(\Theta'_v(\eta_0, v_0)h)(t) = \begin{pmatrix} \Theta'_{1,1} & \Theta'_{1,2} \\ \Theta'_{2,1} & \Theta'_{2,2} \end{pmatrix} \begin{pmatrix} h_1(t) \\ h_2(t) \end{pmatrix} = \begin{pmatrix} 0 \\ 0 \end{pmatrix} \quad \text{for any } t \geq 0,$$

for any $h(t) = \begin{pmatrix} h_1(t) \\ h_2(t) \end{pmatrix} \in \Upsilon$, since the functions G and H have derivatives equal to 0 at zero. For this reason, the operator $(I - \Theta'_v(\eta_0, v_0))^{-1}$ exists and is continuous. According to [168, Theorem 54.2], we conclude that there exist $\delta, \rho > 0$ such that for $\|\eta - \eta_0\|_{P(\sigma^+)E^\beta} \leq \delta$, Eq. (6.4.5) has a unique solution $v(\cdot)$ in the ball $\|v - v_0\|_\Upsilon \leq \rho$ which depends on η continuously. $\square$

Remark 6.3. The results of [263] were obtained under the assumption that $\Sigma_\alpha^+ = \varnothing$. Due to [76], one can extend the results on the existence of stable manifolds and their approximation to the case where the definition of the hyperbolicity of the point u^* in the fractional differential equations case admits the location of the stable part of the spectrum of the operator $A_{u^*}^-$ on the right-hand side of the complex plane (but of course outside of Λ_α^+). The results on the existence of stable manifolds and their approximation in the case where $\Sigma_\alpha^+ \neq \varnothing$ follow the same way as the consideration of Σ_α^+ for the unstable manifold.

6.5 Approximation of Unstable Manifolds

Consider the family of well-posed Cauchy problems in Banach spaces E_n^β:

$$(D_t^\alpha u_n)(t) = A_n u_n(t) + f_n(u_n(t)), \quad t \geq 0,$$
$$u_n(0) = u_n^0 \in E_n^\beta, \tag{6.5.1}$$

where $u_n^0 \xrightarrow{\mathcal{P}^\beta} u^0$, the operators (A_n, A) are compatible, $f_n(\cdot) : E_n^\beta \to E_n$ are globally bounded and globally Lipschitz continuous both uniformly in $n \in \mathbb{N}$, and continuously Fréchet differentiable.

Under the above assumptions. the mild solution $u_n(\cdot)$ of Eq. (6.5.1) is defined for all $t \geq 0$ (see [148, 263]) and given by the formula

$$u_n(t) = E(t^\alpha A_n)u_n^0 + \int_0^t P_\alpha(t - s, A_n)f_n(u_n(s))ds, \, t \geq 0. \tag{6.5.2}$$

Recall, that a hyperbolic equilibrium point u^* is a solution of the equation $Au^* + f(u^*) = 0$, or equivalently $u^* = -A^{-1}f(u^*)$, such that $\sigma(A + f'(u^*)) \cap \partial\Lambda_\alpha^+ = \varnothing$. Since the operator A has a compact resolvent, the operator $A^{-1}f(\cdot)$ is compact.

In the case where $\Delta_{cc} \neq \varnothing$, we conclude that

$$A_n^{-1}f_n(\cdot) \xrightarrow{\mathcal{PP}} A^{-1}f(\cdot) \quad \text{compactly.}$$

From [288] in the terminology of [263] it follows that

$$1 = \gamma(I + A^{-1}f'(u^*); \partial\Omega) = \gamma(I_n + A_n^{-1}f_n(u_n^*); \partial\Omega_n)$$

for each $u^* \in E$ and then the equations $u_n^* = -A_n^{-1}f_n(u_n^*)$ have solutions such that $u_n^* \xrightarrow{\mathcal{P}} u^*$. As was also shown in [120, 235] under the condition $\Delta_{cc} \neq \varnothing$ for each $\lambda \in \Sigma_\alpha^+$ there are points $\lambda_n \in \Sigma_{\alpha,n}^+$ such that $\lambda_n \to \lambda$ and, moreover, projectors $P_n(\Sigma_{\alpha,n}^+) \xrightarrow{\mathcal{PP}} P(\Sigma_\alpha^+)$ compactly, where $\Sigma_{\alpha,n}^+$ are defined as Σ_α^+, but for the operators $A_{u_n^*,n}$.

Consider now the family of the problems (6.5.1) in a neighborhood of hyperbolic equilibrium points u_n^*. In this case, we obtain

$$(\mathbf{D}_t^\alpha v_n)(t) = A_{u_n^*,n}v_n(t) + F_{u_n^*,n}(v_n(t)), \quad t \geq 0,$$
$$v_n(0) = v_n^0, \tag{6.5.3}$$

where

$$A_{u_n^*,n} = A_n + f_n'(u_n^*),$$
$$F_{u_n^*,n}(v_n(t)) = f_n(v_n(t) + u_n^*) - f_n(u_n^*) - f_n'(u_n^*)v_n(t).$$

From now on, we consider a hyperbolic equilibrium u^* of (5.1.1) and hyperbolic equilibria $u_n^* \xrightarrow{\mathcal{P}^\beta} u^*$ of (6.5.1).

We decompose E_n^β using the projections

$$P_n(\sigma_n^+) := P_n(\sigma_n^+, A_{u_n^*,n}) := \frac{1}{2\pi i} \int_{\partial U(\sigma_n^+)} \left(\zeta I_n - A_{u_n^*,n}\right)^{-1} d\zeta \qquad (6.5.4)$$

determined by the sets $\sigma_n^+ \in \Lambda_\alpha^+$, which approximate the set σ^+.

Owing to the analyticity of the C_0-semigroup $e^{tA_{u_n^*,n}}$ and the condition $\Delta_{cc} \neq \varnothing$, there exist some positive $M_2, \gamma > 0$ such that the operator $A_{u_n^*,n}$ satisfied the estimate

$$\left\| e^{tA_{u_n^*,n}} \eta_n \right\|_{E_n^\beta} \leq M_2 e^{-\gamma t} \|\eta_n\|_{E_n^\beta}, \quad t \geq 0, \qquad (6.5.5)$$

for any $\eta_n \in (I_n - P_n(\Sigma_{\alpha,n}^+) - P_n(\sigma^+))E_n^\beta$. The proof of this fact is the same as in [69]. This means that we can use the estimates of [263, Lemma 3.3] on the subspaces $(I_n - P_n(\Sigma_{\alpha,n}^+) - P_n(\sigma^+))E_n^\beta$.

In the same way as we introduced the operator $\Theta(\eta, v(\cdot))$ in (6.4.5), we can define the operator $\Theta_n(\eta_n, v_n(\cdot))$. We assume that $\eta_n \xrightarrow{\mathcal{P}^\theta} \eta$. Rewriting (6.5.1) for $v_n(t) = u_n(t) - u_n^*$ to deal with the neighborhood of u_n^*, we arrive at (6.5.3), which can be rewritten in the following way:

$$D_t^\alpha w_n(t) = (A_n + f_n'(u_n^*))P_n(\sigma_n^+)w_n(t) + H_n(w_n(t), z_n(t)),$$
$$D_t^\alpha z_n(t) = (A_n + f_n'(u_n^*))(I_n - P_n(\sigma_n^+))z_n(t) + G_n(w_n(t), z_n(t)), \qquad (6.5.6)$$

where $w_n(\cdot) \in P_n(\sigma_n^+)E_n^\beta$ and $z_n(\cdot) \in (I_n - P_n(\sigma_n^+))E_n^\beta$.

Theorem 6.11 (see [230]). *Let $\Delta_{cc} \neq \varnothing$ and Conditions (F1)–(F3) be satisfied. Then the operators*

$$\Theta_n(\eta_n, \cdot) : \Upsilon_n \to \Upsilon_n, \quad \Theta(\eta, \cdot) : \Upsilon \to \Upsilon$$

are compact and they converge compactly, $\Theta_n(\eta_n, \cdot) \to \Theta(\eta, \cdot)$, whenever $\eta_n \xrightarrow{\mathcal{P}^\beta} \eta$.

Proof. Recall that one can write for $q_n(t)$, $t \leq 0$, with $q_n(-\infty) = 0$

$$\|q_n(\cdot)\|_{\Upsilon_n} = \left\| \begin{pmatrix} q_n^1(\cdot) \\ q_n^2(\cdot) \end{pmatrix} \right\|_{\Upsilon_n}$$

$$= \sup_{t \leq 0} \|q_n^1(t)\|_{P_n(\sigma_n^+)E_n^\beta} + \sup_{t \leq 0} \|q_n^2(t)\|_{(I_n - P_n(\sigma_n^+))E_n^\beta}.$$

So the discrete convergence $\Upsilon_n \ni q_n(\cdot) \to q(\cdot) \in \Upsilon$ means that

$$\sup_{t \le 0} \left\| q_n^1(t) - p_n^\beta q^1(t) \right\|_{P_n(\sigma_n^+)E_n^\beta} + \sup_{t \le 0} \left\| q_n^2(t) - p_n^\beta q^2(t) \right\|_{(I_n - P_n(\sigma_n^+))E_n^\beta} \to 0$$

as $n \to \infty$. Establishing the compactness of the operators $\Theta_n(\eta_n, \cdot)$ and $\Theta(\eta, \cdot)$ follows the same scheme as the proof of the compact convergence $\Theta_n(\eta_n, \cdot) \to \Theta(\eta, \cdot)$, which we will present. The fact of convergence $\Theta_n(\eta_n, \cdot) \to \Theta(\eta, \cdot)$ is obvious. It follows from the theorem on majorant convergence and the ABC Theorem (see Theorem 1.3 in Chapter. 1).

Now we verify the following condition: $\|q_n(\cdot)\|_{\Upsilon_n} = O(1)$ implies that the sequence $\{\Theta_n(\eta_n, q_n(\cdot))\}$ is P^β-compact. The proof is divided into two steps.

The finite-dimensional part (see $P_n(\sigma_n^+)E_n^\beta$) according to the representation (as in (6.4.4))

$$E_\alpha\big((-t)^\alpha A_{u_n^*,n}^+\big)^{-1} \eta_n - E_\alpha\big((-t)^\alpha A_{u_n^*,n}^+\big)^{-1}$$
$$\int_t^0 (-s)^{\alpha-1} E_{\alpha,\alpha}\big((-s)^\alpha A_{u_n^*,n}^+\big) H_n\big(q_n^1(s), q_n^2(s)\big) ds, \quad t \le 0,$$

is compact since the operators $A_{u_n^*,n}^+ P_n(\sigma_n^+)$ are finite-dimensional, have uniformly in n bounded spectra located in Λ_α^+ that approximate the spectrum of $A_{u_*} P(\sigma^+)$, the families $S_\alpha(t, A_{u_n^*,n}^+), P_\alpha(t, A_{u_n^*,n}^+)$ are uniformly bounded in n by the ABC Theorem, and starting from some large $t_1, t_2 > M$ this part of the solutions becomes equicontinuous. Indeed, formally

$$\left\| P_j \left(S_\alpha(t_1, A_{u_n^*,n}^+)^{-1} \int_0^{t_1} P_\alpha(s, A_{u_n^*,n}^+) H_n(q_n^1(s), q_n^2(s)) ds \right. \right.$$

$$\left. \left. - S_\alpha(t_2, A_{u_n^*,n}^+)^{-1} \int_0^{t_2} P_\alpha(s, A_{u_n^*,n}^+) H_n(q_n^1(s), q_n^2(s)) ds \right) \right\|$$

$$\le \left\| E_\alpha(t_1)^{-1} \int_0^{t_1} P_\alpha \cdots ds - E_\alpha(t_2)^{-1} \int_0^{t_1} P_\alpha \cdots ds \right\|$$

$$+ \left\| E_\alpha(t_2)^{-1} \int_0^{t_1} P_\alpha \cdots ds - E_\alpha(t_2)^{-1} \int_0^{t_2} P_\alpha \cdots ds \right\|$$

$$\le \left\| \big(E_\alpha(t_1)^{-1} - E_\alpha(t_2)^{-1}\big) \int_0^{t_1} P_\alpha \cdots ds \right\|$$

$$+ \left\| E_\alpha(t_2)^{-1} \int_{t_2}^{t_1} P_\alpha \cdots ds \right\|$$

$$\leq \left(e^{-\omega t_1} - e^{-\omega t_2}\right) \int_0^{t_1} \|P_\alpha \cdots ds\| + e^{-\omega t_2}\left(e^{\omega t_1} - e^{\omega t_2}\right) \leq c|t_1 - t_2|.$$

On the interval $[0, M]$, the functions are also equicontinuous. Assume, to the contrary, that there exists $t_0 \in [0, M]$ such that $E_\alpha(\lambda_n t_n^\alpha) \to 0$ as $t_n \to t_0$. Here we have $\lambda_n \to \lambda \in \sigma(A_{u*}^+)$. By the consistency, $E_\alpha(\lambda t_0^\alpha) = 0$, which is impossible since there are no zeros of the function $E_\alpha(\lambda t_0^\alpha)$ in Λ_α^+. By [263, (2.3)], this means that we have uniform continuity in t:

$$\left\| E_\alpha(t_1)^{-1} - E_\alpha(t_2)^{-1} \right\| = \left\| E_\alpha(t_1)^{-1} E_\alpha(t_2)^{-1}\left(E_\alpha(t_1) - E_\alpha(t_2)\right) \right\|$$
$$\simeq c|t_1 - t_2|.$$

On the second step, we show the compactness in the second argument in $(I_n - P_n(\sigma_n^+))E_n^\beta$. This step is split into two parts: the finite-dimensional part $P_n(\Sigma_{\alpha,n}^+)$ and the infinite-dimensional part

$$(I_n - P_n(\Sigma_{\alpha,n}^+) - P_n(\sigma^+))E_n^\beta.$$

For $P_n(\Sigma_{\alpha,n}^+)$, the consideration of

$$(-A_n)^\beta \int_{-\infty}^t P_\alpha(t - s, A_{u*,n}^-)\, G_n(q_n^1(s), q_n^2(s))ds \qquad (6.5.7)$$

is clear as for the finite-dimensional case, since β can be arbitrary and we use the fact that $(-A_n)^\beta = (-A_n)^{-\gamma}(-A_n)^{\beta+\gamma}$ for small $\gamma > 0$. Note that $(-A_n)^{-\gamma} \to (-A)^{-\gamma}$ compactly.

In the case where $(I_n - P_n(\Sigma_{\alpha,n}^+) - P_n(\sigma^+))E_n^\beta$, we can use [263, Lemma 3.3 and Theorem 4.3]. To prove the compact convergence, we apply the Arzelà–Ascoli theorem to the sequence (6.5.7). This sequence is uniformly bounded. Since the integrals converge uniformly in n, for any small $\delta > 0$ the sequence

$$(-A_n)^\beta \int_{-\infty}^{t-\delta} P_\alpha(t - s, A_{u*,n}^-)\, G_n(q_n^1(s), q_n^2(s))ds \qquad (6.5.8)$$

is very close to the previous one. To verify the equicontinuity, we must estimate the difference

$$(-A_n)^\beta \int_{-\infty}^{t+\Delta t-\delta} P_\alpha(t + \Delta t - s, A_{u*,n}^-)\, G_n(q_n^1(s), q_n^2(s))ds$$

$$- (-A_n)^\beta \int_{-\infty}^{t-\delta} P_\alpha(t - s, A_{u*,n}^-)\, G_n(q_n^1(s), q_n^2(s))ds$$

or, formally,

$$(-A_n)^\beta \left(\int_{-\infty}^{t+\Delta t-\delta} P_\alpha(t + \Delta t - s, A_{u^*,n}^-) \cdots ds \right.$$

$$- \int_{-\infty}^{t-\delta} P_\alpha(t + \Delta t - s, A_{u^*,n}^-) \cdots ds$$

$$+ \int_{-\infty}^{t-\delta} P_\alpha(t + \Delta t - s, A_{u^*,n}^-) \cdots ds$$

$$\left. - \int_{-\infty}^{t-\delta} P_\alpha(t - s, A_{u^*,n}^-) \cdots ds \right),$$

which after the change of variables $t - s = \xi$ takes the form

$$(-A_n)^\beta \int_\delta^{\Delta t+\delta} P_\alpha(\xi + \Delta t, A_{u^*,n}^-) \cdots d\xi$$

$$+ (-A_n)^\beta \int_\delta^\infty \left(P_\alpha(\xi + \Delta t, A_{u^*,n}^-) - P_\alpha(\xi, A_{u^*,n}^-) \right) \cdots d\xi.$$

The first term is estimated by Δt. So we must estimate the second term using [263, (2.5)]:

$$(-A_n)^\beta \int_\delta^\infty \left(\int_0^\infty \frac{\tau}{(\xi + \Delta t)^{1-\alpha}} \Phi_\alpha(\tau) e^{\tau(\xi+\Delta t)^\alpha A_{u^*,n}^-} d\tau \right.$$

$$\left. - \int_0^\infty \frac{\tau}{\xi^{1-\alpha}} \Phi_\alpha(\tau) e^{\tau(\xi+\Delta t)^\alpha A_{u^*,n}^-} d\tau \right) \cdots d\xi$$

$$+ (-A_n)^\beta \int_0^\infty \frac{\tau}{\xi^{1-\alpha}} \Phi_\alpha(\tau) \left(e^{\tau(\xi+\Delta t)^\alpha A_{u^*,n}^-} d\tau - e^{\tau\xi^\alpha A_{u^*,n}^-} \right) d\tau \cdots d\xi.$$

Since

$$\left\| \left(-A_{u^*,n}^- \right)^\beta e^{\tau(\xi+\Delta t)^\alpha A_{u^*,n}^-} \right\| \leq \frac{C}{(\tau(\xi + \Delta t)^\alpha)^\beta} e^{-\omega\tau(\xi+\Delta t)^\alpha},$$

the estimate of the first line can be performed by using the expression

$$\frac{1}{(\xi + \Delta t)^{1-\alpha}} - \frac{1}{\xi^{1-\alpha}}$$

and the fact that $\xi \geq \delta$. The second line is modified by the term

$$(-A_{u^*,n}^-)^{1+\beta} \int_{\tau\xi^\alpha}^{\tau(\xi+\Delta t)^\alpha} e^{\eta A_{u^*,n}^-} d\eta$$

and we use an estimate of

$$\frac{1}{(\xi + \Delta t)^{\alpha(1+\beta)}} - \frac{1}{\xi^{\alpha(1+\beta)}}.$$

So we can apply the Arzelà–Ascoli theorem. $\qquad\qquad\square$

Theorem 6.12 (see [230]). *Let $\Delta_{cc} \neq \varnothing$, let Conditions (F1)–(F3) be fulfilled, and let u^* be a hyperbolic equilibrium point of (5.1.1). Then for sufficiently small $\delta > 0$, Eq. (6.4.5) has a solution $\zeta^*(\cdot)$ such that the equations $\zeta_n(t) = \Theta_n(\eta_n, \zeta_n(t))$ have solutions $\zeta_n^*(\cdot)$ for $n \geq n_0$ and $\zeta_n^*(t) \to \zeta^*(t)$ uniformly in $t \leq 0$ as $n \to \infty$.*

Proof. We know that $|\operatorname{ind}(\zeta_0, I - \Theta)| = 1$ for $\zeta_0 = 0$. According to [168, Theorem 54.1], there exist $\rho, \delta > 0$ such that $|\operatorname{ind}(\zeta, I - \Theta)| = 1$ for $\|\eta - 0\|_{E^\beta} \leq \delta$ and $\|\zeta - 0\|_\Upsilon \leq \rho$. By Theorem 6.11, the operator convergence $\Theta_n \to \Theta$ is compact, so

$$\gamma(I_n - \Theta_n, \partial \Omega_n) = \operatorname{ind}(\zeta, I - \Theta)$$

and, therefore, the equation $\zeta_n(t) = \Theta_n(\eta_n, \zeta_n(t))$ has at least one solution $\zeta_n^*(\cdot)$ on any set Ω_n for $n \geq n_0$. Here

$$\Omega_n = \left\{ u_n(\cdot) : \sup_{-\infty < t \leq 0} \left\| u_n(t) - u^* \right\|_{E_n^\beta} \leq \rho \right\}.$$

This sequence $\{\zeta_n^*(\cdot)\}$ is discretely compact and $\zeta_n^*(t) \to \zeta^*(t)$ uniformly in $t \leq 0$ as $n \to \infty$. $\qquad\square$

Theorem 6.13 (see [230]). *Assume also that u^* is a hyperbolic equilibrium for (5.1.1) and that zero is not in the spectrum of the operator $A + f'(u^*) : D(A) \subset E \to E$. Assume that $\Delta_{cc} \neq \varnothing$ and Conditions (F1)–(F3) are fulfilled. Then there exist $\delta > 0$ and $n_0 > 0$ such that u_n^* has a local unstable manifold $W_{loc}^u(u_n^*) \subset E_n^\beta$ for any $n \geq n_0$ and if we denote its δ-part by*

$$W_{n,\delta}^u(u_n^*) = \left\{ v_n(\cdot) \in W_{loc}^u(u_n^*) : \|v_n(t) - u_n^*\|_{E_n^\beta} < \delta, \ t \leq 0 \right\}, \quad n \geq n_0,$$

then $W_{n,\delta}^u(u_n^*)$ converges to $W_\delta^u(u^*)$ as $n \to \infty$, that is,

$$\sup_{v_n \in W_{n,\delta}^u(u_n^*)} \inf_{v \in W_\delta^u(u^*)} \left\| v_n(t) - p_n^\beta v(t) \right\|_{E_n^\beta}$$

$$+ \sup_{v \in W_\delta^u(u^*)} \inf_{v_n \in W_{n,\delta}^u(u_n^*)} \left\| v_n(t) - p_n^\beta v(t) \right\|_{E_n^\beta} \to 0 \quad as \ n \to \infty$$

for any $t \leq 0$.

Proof. We show that for a suitably small $\rho > 0$, there is an unstable manifold for u_n^*

$$W_{n,\delta}^u = \left\{ w_n(\cdot) + z_n(\cdot) : z_n(\cdot) \in (I_n - P_n(\sigma_n^+)) E_n^\beta, \ w_n(\cdot) \in P_n(\sigma_n^+) E_n^\beta \right\}.$$

For the vector

$$\zeta_n(t) = \begin{pmatrix} w_n(t) \\ z_n(t) \end{pmatrix},$$

we see from [263, (4.9)] the following. There is $\delta > 0$ such that the operators $\Theta_n(\eta_n, \cdot)$ are Fréchet differentiable in the balls $\mathcal{U}_{E_n^\beta}(p_n^\beta \zeta^*, \delta)$ and for any $\varepsilon > 0$, there exists $\delta_\varepsilon > 0$ such that

$$\left\| \Theta_n'(\eta_n, \zeta_n) - \Theta_n'(\eta_n, p_n^\alpha \zeta^*) \right\| \le \varepsilon$$

whenever

$$\left\| \zeta_n(\cdot) - p_n^\beta \zeta^*(\cdot) \right\|_{\Upsilon_n} \le \delta_\varepsilon.$$

It is clear also that the conditions

$$\left\| p_n^\beta \zeta^*(\cdot) - \Theta_n(\eta_n, p_n^\beta \zeta^*(\cdot)) \right\|_{\Upsilon_n} \to 0,$$

$$\left\| \left(I_n - \Theta_n'(\eta_n, p_n^\beta \zeta^*(\cdot)) \right)^{-1} \right\| \le \text{const} \quad \text{as } n \to \infty,$$

are satisfied. Then by [288, Theorem 2], it follows that there exist $n_0 \in \mathbb{N}$ and $0 < \delta_0 \le \delta$ such that for each $n \ge n_0$, Eqs. (6.5.6) have unique solutions $\zeta_n^*(\cdot)$ in the balls $\mathcal{U}_{E_n^\beta}(p_n^\beta \zeta^*, \delta_0)$ and $\zeta_n^*(\cdot) \to \zeta^*(\cdot)$ as $n \to \infty$. $\qquad \square$

Bibliography

[1] S. B. Agase and V. Raghavendra, Existence of mild solutions of semilinear differential equations in Banach spaces, *Indian J. Pure Appl. Math.* **21**, 9, pp. 813–821 (1990).

[2] M. Ahues and S. Piskarev, Spectral approximation of weakly singular integral operators. I. Convergence theory, in *Proc. 4 Int. Conf. "Integral Methods in Science and Engineering," Oulu, Finland, June 17–20, 1996.* Longman (1996).

[3] R. R. Akhmerov, M. I. Kamenskii, A. S. Potapov, A. E. Rodkina and B. N. Sadovskii, *Measures of Noncompactness and Condensing Operators.* Birkhäuser-Verlag, Basel (1992).

[4] G. Akrivis, Stability of implicit and implicit-explicit multistep methods for nonlinear parabolic equations, *IMA J. Numer. Anal.* **38**, 4, pp. 1768–1796 (2018).

[5] A. A. Alikhanov, A new difference scheme for the time fractional diffusion equation, *J. Comput. Phys.* **280**, pp. 424–438 (2015).

[6] H. M. Alkhayuon and P. Ashwin, Rate-induced tipping from periodic attractors: partial tipping and connecting orbits, *Chaos* **28**, 3 (2018), paper ID 033608.

[7] F. Alouges and A. Debussche, On the qualitative behavior of orbits of a parabolic partial differential equation and its discretization in the neighborhood of a hyperbolic fixed point, *Numer. Funct. Anal. Optim.* **12**, pp. 253–269 (1991).

[8] E. Alvarez and C. Lizama, Weighted pseudo almost periodic solutions to a class of semilinear integro-differential equations in Banach spaces, *Adv. Differ. Equ.* **2015** (2015), paper ID 31.

[9] H. Amann, Existence and regularity for semilinear parabolic evolutions, *Ann. Scu. Norm. Sup. Pisa Cl. Sci.* **11**, 4, pp. 593–675 (1984).

[10] H. Amann, *Linear and Quasilinear Parabolic Problems. Vol. I. Abstract Linear Theory.* Birkhäuser-Verlag, Basel (1995).

[11] A. Ambrosetti and G. Prodi, *A Primer of Nonlinear Analysis.* Cambridge Univ. Press, Cambridge (1995).

[12] B. Amir and L. Maniar, Existence and asymptotic behavior of solutions of semilinear Cauchy problems with nondense domain via extrapolation spaces, *Rend. Circ. Mat. Palermo, II. Ser.* **49**, 3, pp. 481–496 (2000).

[13] P. M. Anselone, *Collectively Compact Operator Approximation Theory and Applications to Integral Equations.* Prentice-Hall, Englewood Cliffs, New Jersey (1971).

[14] A. V. Antoniouk, A. N. Kochubei and S. I. Piskarev, On the compactness and the uniform continuity of a resolution family for a fractional differential equation, *Dopov. Nats. Akad. Nauk Ukr.*, 6, pp. 7–12 (2014).

[15] R. Aparicio and V. Keyantuo, Well-posedness of degenerate integro-differential equations in function spaces, *Electr. J. Differ. Equ.* **2018** (2018), paper ID 79.

[16] J. Appell, Measures of noncompactness, condensing operators and fixed points: An application-oriented survey, *Fixed Point Theory* **6**, 2, pp. 157–229 (2005).

[17] J. Appell, E. D. Pascale and A. Vignoli, *Nonlinear Spectral Theory.* de Gruyter, Berlin (2004).

[18] J. Appell and P. Zabrejko, *Nonlinear Superposition Operators.* Cambridge Univ. Press, Cambridge (1990).

[19] F. Aràndiga and V. Caselles, On strongly stable approximations, *Rev. Mat. Univ. Complut. Madrid* **7**, 2, pp. 207–217 (1994).

[20] W. Arendt, C. J. Batty, M. K. Hieber and F. Neubrander, *Vector-Valued Laplace Transforms and Cauchy Problems.* Birkhäuser-Verlag, Basel (2001).

[21] J. M. Arrieta, A. N. Carvalho and G. Lozada-Cruz, Dynamics in dumbbell domains. I. continuity of the set of equilibria, *J. Differ. Equ.* **231**, pp. 551–597 (2006).

[22] J. M. Arrieta, A. N. Carvalho and A. Rodriguez-Bernal, Parabolic problems with nonlinear boundary conditions and critical nonlinearities, *J. Differ. Equ.* **156**, 2, pp. 376–406 (1999).

[23] J. M. Arrieta, A. N. Carvalho and A. Rodriguez-Bernal, Attractors of parabolic problems with nonlinear boundary conditions. uniform bounds, *Commun. Partial Differ. Equ.* **25**, 1-2, pp. 1–37 (2000).

[24] J. M. Arrieta, A. N. Carvalho and A. Rodriguez-Bernal, Upper semicontinuity for attractors of parabolic problems with localized large diffusion and nonlinear boundary conditions, *J. Differ. Equ.* **168**, 1, pp. 33–59 (2000).

[25] A. Ashyralyev, A note on fractional derivatives and fractional powers of operators, *J. Math. Anal. Appl.* **357**, pp. 232–236 (2009).

[26] A. Ashyralyev, C. Cuevas and S. Piskarev, On well-posedness of difference schemes for abstract elliptic problems in $L^p([0,T];E)$ spaces, *Numer. Funct. Anal. Optim.* **29**, 1-2, pp. 43–65 (2008).

[27] A. Ashyralyev, M. Martinez, J. Pastor and S. Piskarev, On well-posedness of abstract hyperbolic problems in function spaces, *Proc. WSPC*, pp. 679–688 (2009).

[28] A. Ashyralyev, J. Pastor, S. Piskarev and H. A. Yurtsever, Second-order equations in functional spaces: qualitative and discrete well-posedness, *J. Abstr. Appl. Anal.* (2015), paper ID 948321.

[29] A. Ashyralyev, S. Piskarev and L. Weis, On well-posedness of difference schemes for abstract parabolic equations in $L^p([0,T];E)$ spaces, *Numer. Funct. Anal. Optim.* **23**, 7-8, pp. 669–693 (2002).

[30] A. Ashyralyev and P. E. Sobolevskii, *Well-Posedness of Parabolic Difference Equations*. Birkhäuser-Verlag, Basel–Boston–Berlin (1994).

[31] A. Ashyralyev and P. E. Sobolevskii, *New Difference Schemes for Partial Differential Equations*. Birkhäuser-Verlag, Basel–Boston–Berlin (2004).

[32] A. Atangana, *Derivative with a New Parameter. Theory, Methods and Applications*. Elsevier/Academic Press, Amsterdam (2016).

[33] A. Atangana, *Fractional Operators with Constant and Variable Order with Application to Geo-Hydrology*. Elsevier/Academic Press, Amsterdam (2018).

[34] A. Atangana, Non validity of index law in fractional calculus: A fractional differential operator with Markovian and non-Markovian properties, *Phys. A* **505**, pp. 688–706 (2018).

[35] A. Atangana and E. F. D. Goufo, Conservatory of Kaup–Kupershmidt equation to the concept of fractional derivative with and without singular kernel, *Acta Math. Appl. Sin.* **34**, 2, pp. 351–361 (2018).

[36] Y. I. Babenko, *Method of Fractional Derivatives in Applied Problems of Heat and Mass Transfer*. Saint Petersburg (2009).

[37] E. G. Bajlekova, *Fractional evolution equations in Banach spaces*, Ph.D. thesis, Eindhoven Univ. Technology (2001).

[38] N. Bakaev, V. Thomee and L. B. Wahlbin, Maximum-norm estimates for resolvents of elliptic finite element operators, *Math. Comp.* **72**, pp. 1597–1610 (2003).

[39] N. Y. Bakaev, *Linear Discrete Parabolic Problems*. North-Holland, Amsterdam (2006).

[40] J. Banas and K. Goebel, *Measures of Noncompactness in Banach Spaces*. Marcel Dekker, New York–Basel (1980).

[41] V. Barbu and N. Pavel, On the invertibility of $i+\exp(-ta)$ with a maximal monotone, in *Proc. World Congress in Nonlinear Analysis, Tampa, Florida, August 19–26, 1992*. de Gruyter, pp. 2231–2237 (1992).

[42] A. G. Baskakov, Linear differential operators with unbounded operator coefficients and semigroups of difference operators, *Mat. Zametki* **59**, 6, pp. 811–820 (1996).

[43] A. G. Baskakov, Spectral analysis of linear differential operators and semigroups of difference operators, *Differ. Uravn.* **37**, 1, pp. 3–11 (2001).

[44] A. G. Baskakov, On differential and difference Fredholm operators, *Dokl. Ross. Akad. Nauk* **416**, 2, pp. 156–160 (2007).

[45] A. G. Baskakov, *Study of Differential Equations with Unbounded Operator Coefficients by Methods of the Operator Theory*. Voronezh (2017).

[46] A. G. Baskakov, N. S. Kaluzhina and D. M. Polyakov, Slowly varying on infinity semigroups of operators, *Izv. Vyssh. Ucheb. Zaved. Mat.* **7**, pp. 3–14 (2014).

[47] A. G. Baskakov and I. A. Krishtal, Spectral analysis of operators with the two-point bohr spectrum, *J. Math. Anal. Appl.* **308**, 2, pp. 420–439 (2005).

[48] A. G. Baskakov and A. I. Pastukhov, Spectral analysis of a weighted shift operator with unbounded operator coefficients, *Sib. Mat. Zh.* **42**, 6, pp. 1231–1243 (2001).

[49] A. G. Baskakov and Y. N. Sintyaev, Finite-difference operators in the study of differential operators: solution estimates, *J. Differ. Equ.* **46**, 2, pp. 214–223 (2010).

[50] J. K. Batty Charles, R. Chill and Y. Tomilov, Strong stability of bounded evolution families and semigroups, *J. Funct. Anal.* **193**, 1, pp. 116–139 (2002).

[51] N. C. Bernardes Jr., P. R. Cirilo, U. B. Darji, A. Messaoudi and E. R. Pujals, Expansivity and shadowing in linear dynamics, *J. Math. Anal. Appl.* **461**, 1, pp. 796–816 (2018).

[52] W.-J. Beyn, On the numerical approximation of phase portraits near stationary points, *SIAM J. Numer. Anal.* **24**, 5, pp. 1095–1113 (1987).

[53] W.-J. Beyn, The numerical computation of connecting orbits in dynamical systems, *IMA J. Numer. Anal.* **10**, 3, pp. 379–405 (1990).

[54] W.-J. Beyn, Numerical methods for dynamical systems, in *Proc. 4th Summer School "Advances in Numerical Analysis," Lancaster, UK, 1990.* Oxford Univ. Press, New York, pp. 175–236 (1991).

[55] W.-J. Beyn, V. S. Kolezhuk and S. Y. Pilyugin, Convergence of discretized attractors for parabolic equations on the line, *Zap. Nauch. Semin. LOMI* **318**, pp. 14–41 (2004).

[56] W.-J. Beyn and S. I. Piskarev, Shadowing for discrete approximations of abstract parabolic equations, *Discr. Cont. Dynam. Syst. Ser. B* **10**, 1, pp. 19–42 (2008).

[57] N. A. Bobylev, J. K. Kim, S. K. Korovin and S. I. Piskarev, Semidiscrete approximations of semilinear periodic problems in Banach spaces, *Nonlin. Anal.* **33**, 5, pp. 473–482 (1998).

[58] N. A. Bobylev and M. A. Krasnosel'skii, A functionalization of the parameter and a theorem of relatedness for autonomous systems, *Differ. Uravn.* **6**, 11, pp. 1946–1952 (1970).

[59] N. A. Bobylev and M. A. Krasnosel'skii, On approximation of autorelaxations in systems of automatic control, *Dokl. Akad. Nauk SSSR* **272**, 2, pp. 267–271 (1983).

[60] S. M. Bruschi, A. N. Carvalho, J. W. Cholewa and T. Dlotko, Uniform exponential dichotomy and continuity of attractors for singular perturbed damped wave equation, *J. Dynam. Differ. Equ.* **18**, p. 767 (2006).

[61] S. Bu and G. Cai, Solutions of second-order degenerate integro-differential equations in vector-valued function spaces, *Sci. China Math.* **56**, 5, pp. 1059–1072 (2013).

[62] S. Bu and G. Cai, Well-posedness of second-order degenerate differential equations with finite delay, *Proc. Edinb. Math. Soc.* **60**, pp. 349–360 (2017).

[63] S. Bu and G. Cai, Well-posedness of second-order degenerate differential equations with finite delay in vector-valued function spaces, *Pac. J. Math.* **288**, 1, pp. 27–46 (2017).

[64] S. Bu and G. Cai, Well-posedness of fractional degenerate differential equations with finite delay on vector-valued functional spaces, *Math Nach.* **291**, pp. 759–773 (2018).

[65] E. E. Bukzhalev and A. V. Ovchinnikov, A method of the study of the Cauchy problem for a singularly perturbed linear inhomogeneous differential equation, *Austr. J. Math. Anal. Appl.* **15**, 2, pp.1–14 (2018).

[66] Q. Cao, J. Pastor, S. Siegmund and S. Piskarev, Approximations of parabolic equations at the vicinity of hyperbolic equilibrium point, *Numer. Funct. Anal. Optim.* **35**, 10, pp. 1287–1307 (2014).

[67] V. L. Carbone, A. N. Carvalho and K. Schiabel-Silva, Continuity of attractors for parabolic problems with localized large diffusion, *Nonlin. Anal. Theory Meth. Appl.* **68**, 3, pp. 515–535 (2008).

[68] A. N. Carvalho, J. A. Langa and J. C. Robinson, *Attractors for Infinite-Dimensional Non-Autonomous Dynamical Systems*. Springer-Verlag, Berlin (2013).

[69] A. N. Carvalho and S. I. Piskarev, A general approximation scheme for attractors of abstract parabolic problems, *Numer. Funct. Anal. Optim.* **27**, 7-8, pp. 785–829 (2006).

[70] Y. Chang and R. Ponce, Mild solutions for a multi-term fractional differential equation via resolvent operators, *Mathematics* **6**, 3, pp. 2398–2417 (2021).

[71] F. Chatelin, *Spectral Approximation of Linear Operators*. Academic Press, New York (1983).

[72] J. W. Cholewa and T. Dlotko, *Global Attractors in Abstract Parabolic Problems*. Cambridge Univ. Press, Cambridge (2000).

[73] P. Clément, G. Gripenberg and S.-O. Londen, Regularity properties of solutions of fractional evolution equations, in G. Lumer (ed.), *Evolution Equations and Their Applications in Physical and Life Sciences*. Marcel Dekker, New York, pp. 235–246 (2001).

[74] P. Clément and S.-O. Londen, Regularity aspects of fractional evolution equations, *Rend. Ist. Mat. Univ. Trieste* **31**, 2, pp. 19–20 (2000).

[75] N. D. Cong, T. S. Doan, S. Siegmund and H. T. Tuan, On stable manifolds for planar fractional differential equations, *Appl. Math. Comput.* **226**, pp. 157–168 (2014).

[76] N. D. Cong, T. S. Doan, S. Siegmund and H. T. Tuan, Linearized asymptotic stability for fractional differential equations, *Electron. J. Qual. Theory Differ. Equ.*, 39, pp. 1–13 (2016).

[77] N. D. Cong, T. S. Doan, S. Siegmund and H. T. Tuan, On stable manifolds for fractional differential equations in high-dimensional spaces, *Nonlin. Dynam.* **86**, pp. 1885–1894 (2016).

[78] N. D. Cong, T. S. Doan, S. Siegmund and H. T. Tuan, The instability theorem for nonlinear fractional differential systems, *Discr. Cont. Dyn. Syst. Ser. B* **22**, 8, pp. 3079–3090 (2017).

[79] M. Crouzeix, S. Larsson, S. Piskarev and V. Thomee, The stability of rational approximations of analytic semigroups, *BIT Numer. Math.* **33**, 1, pp. 74–84 (1993).

[80] R. Czaja and P. Marin-Rubio, Pullback exponential attractors for parabolic equations with dynamical boundary conditions, *Taiwan. J. Math.* **21**, 4, pp. 819–839 (2017).

[81] G. da Prato and P. Grisvard, Sommes d'opérateus linéaires et équations différentielles opérationnelles, *J. Math. Pures Appl. IX. Sér.* **54**, 3, pp. 305–387 (1975).

[82] G. da Prato and P. Grisvard, équations d'évolution abstraites non linéaires de type parabolique, *C. R. Acad. Sci. Paris. Sér. A* **283**, 9, pp. 709–711 (1976).

[83] M.-F. Danca, M. Feckan, N. V. Kuznetsov and G. Chen, Complex dynamics, hidden attractors and continuous approximation of a fractional-order hyperchaotic pwc system, *Nonlin. Dynam.* **91**, 4, pp. 2523–2540 (2018).

[84] E. N. Dancer, Upper and lower stability and index theory for positive mappings and applications, *Nonlin. Anal. Theory Meth. Appl.* **17**, pp. 205–217 (1991).

[85] B. de Andrade, A. N. Carvalho, P. M. Carvalho-Neto, and P. Marin-Rubio, Semilinear fractional differential equations: Global solutions, critical nonlinearities and comparison results, *Topol. Meth. Nonlin. Anal.* **45**, 2, pp. 439–467 (2015).

[86] A. Deshpande and V. Daftardar-Gejji, Local stable manifold theorem for fractional systems, *Nonlin. Dynam.* **83**, pp. 2435–2452 (2016).

[87] K. Diethelem, *The Analysis of Fractional Differential Equations*. Springer-Verlag, Heidelberg–Dordrecht–London–New York (2004).

[88] B. Eberhardt and G. Greiner, Baillon's theorem on maximal regularity, *Acta Appl. Math.* **27**, pp. 47–54 (1992).

[89] M. Edelman, Fractional maps and fractional attractors. Part I: α-Families of maps, *Discontin. Nonlinearity Complex.* **1**, 4, pp. 305–324 (2012).

[90] M. Edelman, Caputo standard α-family of maps: Fractional difference vs. fractional, *Chaos* **24**, 2 (2014), paper ID 023137.

[91] M. Edelman, Dynamics of nonlinear systems with power-law memory, in V. E. Tarasov (ed.), *Handbook of Fractional Calculus with Applications*, Vol. 4. de Gruyter, Berlin, pp. 103–132 (2019).

[92] M. M. El-Borai, Some probability densities and fundamental solutions of fractional evolution equations, *Chaos Solitons Fract.* **14**, pp. 433–440 (2002).

[93] K.-J. Engel and R. Nagel, *One-Parameter Semigroups for Linear Evolution Equations*. Springer-Verlag, Berlin (2000).

[94] K.-J. Engel and R. Nagel, *A Short Course on Operator Semigroups*. Springer-Verlag, New York (2006).

[95] T. Ergenç, B. Karasözen and S. Piskarev, Approximation for semilinear Cauchy problems involving second order equations in separable Banach spaces, *Nonlin. Anal. Theory Meth. Appl.* **28**, pp. 1157–1165 (1997).

[96] Z. Fan, Characterization of compactness for resolvents and its applications, *Appl. Math. Comput.* **232**, pp. 60–67 (2014).

[97] Z. Fan, Q. Dong and G. Li, Almost exponential stability and exponential stability of resolvent operator families, *Semigroup Forum* **93**, pp. 491–500 (2016).

[98] V. E. Fedorov and E. A. Romanova, Inhomogeneous fractional evolutionary equation in the sectorial case, *J. Math. Sci.* **250**, 5, pp. 819–829 (2020).

[99] F.Liu and H. X. Wu, Regularity of discrete multisublinear fractional maximal functions, *Sci. China Math.* **60**, pp. 1461–1476 (2017).

[100] H. Fujita and A. Mizutani, On the finite element method for parabolic equations. I. approximation of holomorphic semigroups, *J. Math. Soc. Jpn.* **28**, 4, pp. 749–771 (1976).

[101] H. Fujita and T. Suzuki, Evolution problems, in *Handbook of Numerical Analysis*. North-Holland, Amsterdam, pp. 789–928 (1991).

[102] C. G. Gal and M. Warma, *Fractional-in-Time Semilinear Parabolic Equations and Applications*. Springer (2020).

[103] G. Gao and Z. Sun, The finite difference approximation for a class of fractional sub-diffusion equations on a space unbounded domain, *J. Comput. Phys.* **236**, pp. 443–460 (2013).

[104] B. M. Garay, On structural stability of ordinary differential equations with respect to discretization methods, *Numer. Math.* **72**, 4, pp. 449–479 (1996).

[105] B. M. Garay and K. Lee, Attractors under discretization with variable stepsize, *Discr. Contin. Dynam. Syst.* **13**, 3, pp. 827–841 (2005).

[106] L. Gearhart, Spectral theory for contraction semigroups on Hilbert spaces, *Trans. Am. Math. Soc.* **236**, pp. 385–394 (1978).

[107] I. T. Gohberg and M. G. Krein, *Introduction to the Theory of Linear Non-Self-Aadjoint Operators*. Am. Math. Soc., Providence, Rhode Island (1969).

[108] J. A. Goldstein, *Semigroups of Linear Operators and Applications*. Clarendon Press, Oxford University Press, New York (1985).

[109] R. Gorenflo, A. A. Kilbas, F. Mainardi and S. V. Rogosin, *Mittag-Leffler Functions, Related Topics and Applications*. Springer-Verlag, Berlin (2020).

[110] A. Y. Goritskii and V. V. Chepyzhov, Dichotomy property of solutions of quasilinear equations in problems on inertial manifolds, *Sb. Math.* **196**, 4, pp. 485–511 (2005).

[111] E. F. D. Goufo and J. J. Nieto, Attractors for fractional differential problems of transition to turbulent flows, *J. Comput. Appl. Math.* **339**, pp. 329–342 (2018).

[112] R. D. Grigorieff, Zur Theorie linearer approximationsregulärer Operatoren, I, II, *Math. Nachr.* **55**, pp. 233–249, 251–263 (1973).

[113] R. D. Grigorieff, Diskrete Approximation von Eigenwertproblemen. I. Qualitative Konvergenz, *Numer. Math.* **24**, pp. 355–374 (1975).

[114] R. D. Grigorieff, Diskrete Approximation von Eigenwertproblemen. II. Konvergenzordnung, *Numer. Math.* **24**, pp. 415–433 (1975).

[115] R. D. Grigorieff, Über diskrete Approximationen nichtlinearer Gleichungen. 1. Art, *Math. Nachr.* **69**, pp. 253–272 (1975).

[116] R. D. Grigorieff, Diskrete Approximation von Eigenwertproblemen. III. Asymptotische Entwicklungen, *Numer. Math.* **25**, 1, pp. 79–97 (1975/76).

[117] R. D. Grigorieff, Zur charakterisierung linearer approximationsregulärer operatoren, in *Mathematical Papers Given on the Occasion of Ernst Mohr's 75th Birthday*. Tech. Univ. Berlin, Berlin, pp. 63–77 (1985).

[118] R. D. Grigorieff and H. Jeggle, Approximation von Eigenwertproblemen bei nichlinearer Parameterabhängigkeit, *Manuscr. Math.* **10**, pp. 245–271 (1973).

[119] D. Guidetti, *Fractional derivative in spaces of continuous functions.* preprint (2019).

[120] D. Guidetti, B. Karasözen and S. Piskarev, Approximation of abstract differential equations, *J. Math. Sci.* **122**, 2, pp. 3013–3054 (2004).

[121] I. N. Gurova, On a certain topological method for studying difference schemes, *Dokl. Akad. Nauk SSSR* **248**, 1, pp. 25–28 (1979).

[122] I. N. Gurova, On semidiscretization methods for quasilinear equations with a noncompact semigroup, *Izv. Vyssh. Uchebn. Zaved. Mat.*, 4, pp. 60–65 (2000).

[123] I. N. Gurova and M. I. Kamenskii, On the semidiscretization method in the problem of periodic solutions of quasilinear autonomous parabolic equations, *Differ. Equ.* **32**, 1, pp. 106–112 (1996).

[124] M. Haase, *The Functional Calculus for Sectorial Operators.* Birkhäuser-Verlag, Basel (2006).

[125] J. K. Hale, *Asymptotic Behavior of Dissipative Systems.* Am. Math. Soc. (1989).

[126] J. K. Hale, X. B. Lin and G. Raugel, Upper semicontinuity of attractors for approximations of semigroups and partial differential equations, *Math. Comp.* **50**, pp. 89–123 (1988).

[127] X. Han and P. Kloeden, *Attractors Under Discretisation.* Springer (2017).

[128] B. He, J. Cao and B. Yang, Weighted Stepanov-like pseudo-almost automorphic mild solutions for semilinear fractional differential equations, *Adv. Differ. Equ.* **2015** (2015), paper ID 74.

[129] D. Henry, *Geometric Theory of Semilinear Parabolic Equations.* Springer-Verlag, Berlin (1981).

[130] P. Hess, *Periodic-Parabolic Boundary-Value Problems and Positivity.* Longman (1991).

[131] R. H. W. Hoppe, A constructive approach to the bellman semigroup, *Nonlin. Anal. Theory Meth. Appl.* **9**, 11, pp. 1165–1181 (1985).

[132] Y. Hu and J. H. He, On fractal space-time and fractional calculus, *Thermal Sci.* **20**, 3, pp. 773–777 (2016).

[133] J. Huang, J. Li and T. Shen, Dynamics of stochastic modified boussinesq approximation equation driven by fractional brownian motion, *Dynam. Partial Differ. Equ.* **11**, 2, pp. 183–209 (2014).

[134] M. C. Irwin, *Smooth Dynamical Systems.* World Scientific, River Edge, New Jersey (2001).

[135] B. Jin, R. Lazarov, J. Pasciak and Z. Zhou, Galerkin fem for fractional order parabolic equations with initial data in h^{-s}, $0 \le s \le 1$, in *Numerical Analysis and Its Applications*, pp. 24–37 (2013).

[136] B. Jin, R. Lazarov and Z. Zhou, Error estimates for a semidiscrete finite element method for fractional order parabolic equations, *SIAM J. Numer. Anal.* **51**, 1, pp. 445–466 (2013).

[137] B. Jin, B. Li and Z. Zhou, Discrete maximal regularity of time-stepping schemes for fractional evolution equations, *Numer. Math.* **138**, 1, pp. 101–131 (2018).

[138] M. A. Kaashoek and S. M. Verduyn, An integrability condition on the resolvent for hyperbolicity of the semigroup, *J. Differ. Equ.* **112**, 2, pp. 374–406 (1994).

[139] M. Kamenskii, V. Obukhovskii, G. Petrosyan and Y. Jen-Chih, On approximate solutions for a class of semilinear fractional-order differential equations in Banach spaces, *Fixed Point Theory Appl.* **2017** (2017), paper ID 28.

[140] M. Kamenskii, V. Obukhovskii, G. Petrosyan and Y. Jen-Chih, On approximate solutions for a class of semilinear fractional-order differential equations in Banach spaces, *Fixed Point Theory Appl.* **2017** (2017), paper ID 28.

[141] M. Kamenskii, V. Obukhovskii, G. Petrosyan and Y. Jen-Chih, On semilinear fractional-order differential inclusions in Banach spaces, *Fixed Point Theory* **18**, 1, pp. 269–292 (2017).

[142] M. Kamenskii, V. Obukhovskii, G. Petrosyan and Y. Jen-Chih, Existence and approximation of solutions to nonlocal boundary-value problems for fractional differential inclusions, *Fixed Point Theory Appl.* **2019** (2019), paper ID 2.

[143] M. Kamenskii, V. Obukhovskii and P. Zecca, *Condensing Multivalued Maps and Semilinear Differential Inclusions in Banach Spaces*. De Gruyter, Berlin (2001).

[144] I. Karatay and S. R. Bayramoglu, A characteristic difference scheme for time-fractional heat equations based on the Crank–Nicholson difference schemes, *Abstr. Appl. Anal.* **2012** (2012), paper ID 548292.

[145] S. Katayama, *Global Solutions and the Asymptotic Behavior for Nonlinear Wave Equations with Small Initial Data*. Math. Soc. Jpn., Tokyo (2017).

[146] T. Kato, *Perturbation Theory for Linear Operators*. Springer-Verlag, Berlin (1995).

[147] A. Katok and B. Hasselblatt, *Introduction to the Modern Theory of Dynamical Systems*. Cambridge Univ. Press, Cambridge (1995).

[148] V. Keyantuo, C. Lizama and M. Warma, Spectral criteria for slvability of boundary value problems and positivity of solutions of time-fractional differential equations, *Abstr. Appl. Anal.* **2013** (2013), paper ID 614328.

[149] V. Keyantuo, C. Lizama and M. Warma, Existence, regularity, and representation of solutions of time fractional diffusion equations, *Adv. Differ. Equ.* **21**, 9-10, pp. 837–886 (2016).

[150] V. Keyantuo, C. Lizama and M. Warma, Existence, regularity, and representation of solutions of time fractional wave equations, *Electr. J. Differ. Equ.* **2017** (2017), paper ID 222.

[151] A. A. Kilbas, H. M. Srivastava and J. J. Trujillo, *Theory and Applications of Fractional Differential Equations.* Elsevier, Amsterdam (2006).

[152] P. E. Kloeden, Upper semicontinuity of attractors of delay differential equations in the delay, *Bull. Austr. Math. Soc.* **73**, pp. 299–305 (2006).

[153] P. E. Kloeden and J. Lorenz, Stable attracting sets in dynamical system and their one-step discretizations, *SIAM J. Numer. Anal.* **23**, pp. 986–995 (1986).

[154] P. E. Kloeden and J. Lorenz, Lyapunov stability and attractors under discretization, in *Proc. Equadiff 87 Conf., Xanthi, Greece, August 1987.* Marcel Dekker, Paris, pp. 361–368 (1989).

[155] P. E. Kloeden and S. I. Piskarev, Discrete convergence and the equivalence of equi-attraction and the continuous convergence of attractors, *Int. J. Dynam. Syst. Differ. Equ.* **1**, pp. 38–43 (2007).

[156] M. M. Kokurin, The uniqueness of a solution to the inverse Cauchy problem for a fractional differential equation in a Banach space, *Izv. Vyssh. Uchebn. Zaved. Mat.*, 12, pp. 19–35 (2013).

[157] M. M. Kokurin and S. I. Piskarev, A finite difference scheme on a graded mesh for solving Cauchy problems with a fractional Caputo derivative in a Banach space, *Izv. Vyssh. Ucheb. Zaved. Mat.* **11**, pp. 38–51 (2022).

[158] L. A. Kondratieva and A. V. Romanov, Inertial manifolds and limit cycles of dynamical systems in $\mathbb{R}^n$, *Electron. J. Qual. Theory Differ. Equ.* **2019** (2019), paper ID 96.

[159] A. A. Kornev, Approximation of attractors of semidynamical systems, *Sb. Math.* **192**, 10, pp. 1435–1450 (2001).

[160] M. O. Korpusov and A. V. Ovchinnikov, *Blow-Up of Solutions of Model Nonlinear Equations of Mathematical Physics.* URSS, Moscow (2013).

[161] M. O. Korpusov and A. V. Ovchinnikov, *Blow-Up of Solutions to Nonlinear Equations and Systems of Equations of Mathematical Physics.* URSS, Moscow (2017).

[162] M. O. Korpusov, A. V. Ovchinnikov, A. G. Sveshnikov and E. V. Yushkov, *Blow-Up in Nonlinear Equations of Mathematical Physics. Theory and Methods.* De Gruyter, Berlin (2018).

[163] M. O. Korpusov, A. V. Ovchinnikov and A. A. Panin, Instantaneous blow-up versus local solvability of solutions to the Cauchy problem for the equation of a semiconductor in a magnetic field, *Math. Meth. Appl.* **41**, pp. 8070–8099 (2018).

[164] M. O. Korpusov, A. V. Ovchinnikov and A. A. Panin, *Nonlinear Functional Analysis.* World Scientific, Singapore (2022).

[165] M. Kostić, A note on semilinear fractional equations governed by abstract differential operators, *An. Ştiinţ. Univ. ACuza Iaşi l. I., Ser. Nouă, Mat.* **62**, 2, pp. 757–762 (2016).

[166] I. N. Kostin, Lower semicontinuity of a nonhyperbolic attractor, *J. London Math. Soc.* **52**, pp. 568–582 (1995).

[167] B. Kovacs, B. Li and C. Lubich, A-stable time discretizations preserve maximal parabolic regularity, *SIAM J. Numer. Anal.* **54**, 6, pp. 3600–3624 (2016).

[168] M. A. Krasnosel'skii and P. P. Zabreiko, *Geometrical Methods of Nonlinear Analysis*. Springer-Verlag, Berlin (1984).

[169] M. A. Krasnosel'skyi, P. P. Zabreiko, E. I. Pustyl'nik and P. E. Sobolevskii, *Integral Operators in Spaces of Summable Functions*. Noordhoff, Leiden (1976).

[170] S. G. Krein, *Linear Differential Equations in Banach Space*. Am. Math. Soc., Providence, Rhode Island (1971).

[171] O. Ladyzhenskaya, *Attractors for Semigroups and Evolution Equations*. Cambridge Univ. Press, Cambridge (1991).

[172] S. Larsson, *Nonsmooth data error estimates with applications to the study of the long-time behavior of finite element solutions of semilinear parabolic problems*. Chalmers Univ. Technol. (1992).

[173] S. Larsson, Numerical analysis of semilinear parabolic problems, in M. Ainsworth (ed.), *The Graduate Student's Guide to Numerical Analysis'98*. Springer-Verlag, Berlin, pp. 83–117 (1999).

[174] S. Larsson and S. Y. Pilyugin, *Numerical shadowing near the global attractor for a semilinear parabolic equation*. Chalmers Univ. Technol. (1998).

[175] S. Larsson and J. M. Sanz-Serna, The behavior of finite element solutions of semilinear parabolic problems near stationary points, *SIAM J. Numer. Anal.* **31**, 4, pp. 1000–1018 (1994).

[176] S. Larsson and J.-M. Sanz-Serna, A shadowing result with applications to finite element approximation of reaction-diffusion equations, *Math. Comp.* **68**, 225, pp. 55–72 (1999).

[177] Y. Latushkin, A. Pogan and R. Schnaubelt, Dichotomy and Fredholm properties of evolution equations, *J. Oper. Theory* **58**, 2, pp. 387–414 (2007).

[178] Y. Latushkin, J. Prüss and R. Schnaubelt, Stable and unstable manifolds for quasilinear parabolic systems with fully nonlinear boundary conditions, *J. Evol. Equations* **6**, pp. 537–576 (2006).

[179] Y. Latushkin, J. Prüss and R. Schnaubelt, Center manifolds and dynamics near equilibria of quasilinear parabolic systems with fully nonlinear boundary conditions, *Discr. Cont. Dynam. Syst. Ser. B* **9**, pp. 595–633 (2008).

[180] B. Li, T. Wang and X. Xie, Numerical analysis of a semilinear fractional diffusion equation (2019), arXiv:1909.00016 [math.NA].

[181] C. Li and A. Chen, Numerical methods for fractional partial differential equations, *Int. J. Comp. Math.* **95**, 6-7, pp. 1048–1099 (2018).

[182] C. Y. Li and M. Li, Hölder regularity for abstract fractional Cauchy problems with order in $(0, 1)$, *J. Appl. Math. Phys.* **6**, pp. 310–319 (2018).

[183] D. Li and P. E. Kloeden, Equi-attraction and the continuous dependence of attractors on parameters, *Glasgow Math. J.* **46**, pp. 131–141 (2004).

[184] K. Li and J. Jia, Existence and uniqueness of mild solutions for abstract delay fractional differential equations, *Comput. Math. Appl.* **62**, 3, pp. 1398–1404 (2011).

[185] M. Li, C. Chen and F. Li, On fractional powers of generators of fractional resolvent families, *J. Funct. Anal.* **259**, pp. 2702–2726 (2010).

[186] M. Li, V. Morozov and S. Piskarev, On the approximations of derivatives of integrated semigroups, *J. Inv. Ill-Posed Probl.* **18**, 5, pp. 515–550 (2010).

[187] M. Li, V. Morozov and S. Piskarev, On approximations of derivatives of integrated semigroups, ii, *J. Inv. Ill-Posed Probl.* **19**, 3-4, pp. 643–688 (2011).

[188] M. Li and S. Piskarev, On approximation of integrated semigroups, *Taiwan. J. Math.* **14**, 6, pp. 2137–2161 (2010).

[189] M. Li and Q. Zheng, On spectral inclusions and approximations of a times resolvent families, *Semigroup Forum* **69**, 3, pp. 356–368 (2004).

[190] K.-B. Lin, H. Liu and C.-T. Pang, Approximate controllability of semilinear functional differential equations with hille–yosida operator, *J. Nonlin. Convex Anal.* **17**, 2, pp. 275–285 (2016).

[191] Y. Lin and C. Xu, Finite difference spectral approximations for the time-fractional diffusion equation, *J. Comput. Phys.* **225**, pp. 1533–1552 (2007).

[192] H. Liu and J. C. Chang, Existence for a class of partial differential equations with nonlocal conditions, *Nonlin. Anal. Theory Meth. Appl.* **70**, pp. 3076–3083 (2009).

[193] L. Liu, Z. Fan, G. Li and S. Piskarev, Maximal regularity for fractional Cauchy equation in Hölder space and its approximation, *Comput. Meth. Appl. Math.* **19**, 2, pp. 160–178 (2019).

[194] L. Liu, Z. Fan, G. Li and S. Piskarev, Convergence rates of a finite difference method for the fractional subdiffusion equations, *Adv. Appl. Math. Mech.* (submitted).

[195] R. Liu, M. Li, J. Pastor and S. Piskarev, On the approximation of fractional resolution families, *Differ. Equ.* **50**, 7, pp. 927–937 (2014).

[196] R. Liu, M. Li and S. I. Piskarev, Approximation of semilinear fractional Cauchy problem, *Comput. Meth. Appl. Math.* **15**, 2, pp. 203–212 (2015).

[197] R. Liu, M. Li and S. I. Piskarev, Stability of difference schemes for fractional equations, *Differ. Equ.* **51**, 7, pp. 904–924 (2015).

[198] R. Liu, M. Li and S. I. Piskarev, The order of convergence of difference schemes for fractional equations, *Numer. Funct. Anal. Optim.* **38**, pp. 754–769 (2017).

[199] R. Liu and S. Piskarev, Approximation of semilinear fractional Cauchy problem, II, *Semigroup Forum* **101**, 3, pp. 751–768 (2020).

[200] R. Liu and S. Piskarev, Well-posedness and approximation for nonhomogeneous fractional differential equations, *Numer. Funct. Anal. Optim.* **42**, 6, pp. 619–643 (2021).

[201] C. Lizama, Abstract linear fractional evolution equations, I, in A. Kochubei and Y. Luchko (eds.), *Fractional Differential Equations*. de Gruyter, Berlin–Boston, pp. 465–498 (2019).

[202] C. Lizama, Abstract linear fractional evolution equations, II, in A. Kochubei and Y. Luchko (eds.), *Fractional Differential Equations*. de Gruyter, Berlin–Boston, pp. 499–514 (2019).

[203] C. Lizama, A. Pereira and R. Ponce, On the compactness of fractional resolvent operator functions, *Semigroup Forum* **93**, pp. 363–374 (2016).

[204] G. J. Lord and A. M. Stuart, Discrete gevrey regularity attractors and upper semicontinuity for a finite difference approximation to the Ginzburg–Landau equation, *Numer. Funct. Anal. Optim.* **16**, 7-8, pp. 1003–1047 (1995).

[205] T. Lorenz, Differential equations for closed sets in a Banach space, survey and extension, *Vietnam J. Math.* **45**, 1-2, pp. 5–49 (2017).

[206] F. Mainardi, Fractional diffusive waves in viscoelastic solids, in J. L. Wegner and F. R. Norwood (eds.), *Nonlinear Waves in Solids*. Fairfield, pp. 93–97 (1995).

[207] C. Martínez and M. Sanz, *The Theory of Fractional Powers of Operators*. North-Holland, Amsterdam (2001).

[208] T. Matsumoto and N. Tanaka, Abstract Cauchy problem for weakly continuous operators, *J. Math. Anal. Appl.* **435**, 1, pp. 267–285 (2016).

[209] F. Meng, C. Liu and C. Zhang, Existence of multiple equilibrium points in global attractor for damped wave equation, *Boundary-Value Probl.* **2019** (2019), paper ID 6.

[210] K. S. Miller and B. Ross, *An Introduction to the Fractional Calculus and Fractional Differential Equations*. Wiley, New York (1993).

[211] R. Miyazaki, D. Kim, T. Naito and J. S. Shin, Fredholm operators, evolution semigroups, and periodic solutions of nonlinear periodic systems, *J. Differ. Equ.* **257**, 11, pp. 4214–4247 (2014).

[212] M. Moshrefi-Torbati and J. K. Hammond, Physical and geometrical interpretation of fractional operators, *J. Franklin Inst.* **335B**, 6, pp. 1077–1086 (1998).

[213] A. M. Nakhushev, *Fractional Calculus and Applications* [in Russian] Fizmatlit, Moscow (2003).

[214] O. Nevanlinna and G. Vainikko, Limit spectrum of discretely converging operators, *Numer. Funct. Anal. Optim.* **17**, 7-8, pp. 797–808 (1996).

[215] L. Olszowy, Existence of mild solutions for semilinear nonlocal Cauchy problems in separable Banach spaces, *Z. Anal. Anwend.* **32**, 2, pp. 215–232 (2013).

[216] L. Olszowy, Existence of mild solutions for the semilinear nonlocal problem in Banach spaces, *Nonlin. Anal.* **81**, pp. 211–223 (2013).

[217] A. Ostermann and C. Palencia, Shadowing for nonautonomous parabolic problems with applications to long-time error bounds, *SIAM J. Numer. Anal.* **37**, 5, pp. 1399–1419 (2000).

[218] K. M. Owolabi and A. Atangana, Numerical simulations of chaotic and complex spatiotemporal patterns in fractional reaction-diffusion systems, *Comput. Appl. Math.* **37**, 2, pp. 2166–2189 (2018).

[219] K. M. Owolabi and A. Atangana, Robustness of fractional difference schemes via the Caputo subdiffusion-reaction equations, *Chaos Solitons Fract.* **111**, pp. 119–127 (2018).

[220] K. J. Palmer, A perturbation theorem for exponential dichotomies, *Proc. Roy. Soc. Edinburgh Sec. A* **106**, 1-2, pp. 25–37 (1987).

[221] K. J. Palmer, Exponential dichotomies and Fredholm operators, *Proc. Am. Math. Soc.* **104**, 1, pp. 149–156 (1988).

[222] K. J. Palmer, *Shadowing in Dynamical Systems. Theory and Applications*. Kluwer Academic, Dordrecht (2000).

[223] J. Pastor and S. Piskarev, The exponential dichotomy under discretization on general approximation scheme, *Adv. Numer. Anal.* **2011** (2011), paper ID 582740.

[224] L. Paunonen, Robustness of strong stability of semigroups, *J. Differ. Equ.* **257**, 12, pp. 4403–4436 (2014).

[225] A. Pazy, *Semigroups of Differential Equations.* Marcel Dekker, New York–Basel (1983).

[226] W. V. Petryshyn, *Approximation-Solvability of Nonlinear Functional and Differential Equations.* Marcel Dekker, New York–Basel–Hong Kong (1993).

[227] S. Y. Pilyugin, *Shadowing in Dynamical Systems.* Springer-Verlag, Berlin (1999).

[228] S. Piskarev, On approximation of holomorphic semigroups, *Tartu Riikl. Ul. Toimetised* **492**, pp. 3–23 (1979).

[229] S. Piskarev and S. Siegmund, Approximations of stable manifolds in the vicinity of hyperbolic equilibrium points for fractional differential equations, *Nonlin. Dynam.* **95**, 1, pp. 685–697 (19).

[230] S. Piskarev and S. Siegmund, Unstable manifolds for fractional differential equations, *Eurasian J. Math. Comp. Appl.* **10**, 3, pp. 58–72 (2022).

[231] S. I. Piskarev, Error estimates in the approximation of semigroups of operators by Padé fractions, *Izv. Vyssh. Uchebn. Zaved. Ser. Mat.*, 4, pp. 33–38 (1979).

[232] S. I. Piskarev, Estimates for the rate of convergence in the solution of ill-posed problems for evolution equations, *Izv. Akad. Nauk SSSR. Ser. Mat.* **51**, 3, pp. 676–687 (1987).

[233] S. I. Piskarev, Convergence of difference schemes for solving nonlinear parabolic equations, *Math. Notes* **44**, 1, pp. 549–556 (1988).

[234] S. I. Piskarev, Approximation of positive c_0-semigroups of operators, *Differ. Uravn.* **27**, 7, pp. 1245–1250 (1991).

[235] S. I. Piskarev, *Differential Equations in Banach Spaces and Their Approximation.* Moscow State Univ., Moscow (2005).

[236] I. Podlubny, *Fractional Differential Equations.* Academic Press (1998).

[237] R. Ponce, Existence of mild solutions to nonlocal fractional Cauchy problems via compactness, *Abstr. Appl. Anal.* **2016** (2016), paper ID 4567092.

[238] R. Poongodi and R. Murugesu, Existence of mild solution for fractional nonlocal neutral impulsive integro-differential equations with state-dependent delay, *Nonlin. Stud.* **23**, 2, pp. 209–223 (2016).

[239] R. Poongodi, R. Murugesu and R. Nirmalkumar, Exact controllability results for a class of abstract nonlocal Cauchy problem with impulsive conditions, *Evol. Equ. Control Theory* **6**, 4, pp. 599–613 (2017).

[240] A. Y. Popov and A. M. Sedletskii, Distribution of roots of Mittag-Leffler functions, *J. Math. Sci.* **190**, 2, pp. 209–409 (2013).

[241] J. Prüss, *Evolutionary Integral Equations and Applications.* Birkhäuser-Verlag, Basel (1993).

[242] Y. Ran and J. Li, Pullback attractors for non-autonomous reaction-diffusion equation with infinite delays in $C_{\gamma, L^r(\Omega)}$ or $C_{\gamma, W^{1,r}(\Omega)}$, *Boundary Value Probl.* **2018** (2018), paper ID 99.

[243] R. T. Rau, Hyperbolic evolution groups and dichotomic evolution families, *J. Dynam. Differ. Equ.* **6**, 2, pp. 335–350 (1994).

[244] R. T. Rau, Hyperbolic evolution semigroups on vector valued function spaces, *Semigroup Forum* **48**, 1, pp. 107–118 (1994).

[245] J. C. Robinson, *Infinite-Dimensional Dynamical Systems. An Introduction to Dissipative Parabolic PDEs and the Theory of Global Attractors.* Cambridge Univ. Press, Cambridge (2001).

[246] R. C. Robinson, *An Introduction to Dynamical Systems: Continuous and Discrete.* Pearson Prentice Hall, Upper Saddle River, New Jersey (2004).

[247] S. Rogosin and M. Dubatovskaya, Letnikov vs. marchaud: A survey on two prominent constructions of fractional derivatives, *Mathematics,* **6**, 1 (2018), paper ID 3.

[248] A. V. Romanov, Conditions for asymptotic k-dimensionality of semilinear parabolic equations, *Russ. Math. Surv.* **46**, 1, pp. 255–256 (1991).

[249] A. V. Romanov, Sharp estimates for the dimension of inertial manifolds for nonlinear parabolic equations, *Izv. Ross. Akad. Nauk Ser. Mat.* **57**, 4, pp. 36–54 (1993).

[250] A. V. Romanov, On the limit dynamics of evolution equations, *Usp. Mat. Nauk* **51**, 2, pp. 173–174 (1996).

[251] A. V. Romanov, Finite-dimensional limiting dynamics for dissipative parabolic equations, *Sb. Math.* **191**, 3, pp. 415–429 (2000).

[252] A. V. Romanov, Three counterexamples in the theory of inertial manifolds, *Mat. Zametki* **68**, 3, pp. 439–447 (2000).

[253] A. V. Romanov, On the hyperbolicity properties of inertial manifolds of reaction-diffusion equations (2016), arXiv:1602.08953v1 [math.DS].

[254] A. V. Romanov, On the hyperbolicity properties of inertial manifolds of reaction-diffusion equations, *Dynamics of Partial Differential Equations* **13**, 3, pp. 263–272 (2016).

[255] A. V. Romanov, A parabolic equation with nonlocal diffusion without a smooth inertial manifold, *Math. Notes* **96**, 3-4, pp. 548–555 (2016).

[256] J. Rottmann, *Spectral Properties of Mixed Hyperbolic-Parabolic Systems.* Univ. Bielefeld (2005).

[257] S. Samko, A. Kilbas and O. Marichev, *Fraction Integrals and Derivatives.* Gordon and Breach, New York (1993).

[258] H. Sano and N. Tanaka, Well-posedness and flow invariance for semilinear functional differential equations governed by nondensely defined operators, *Differ. Integral Equations* **30**, 9-10, pp. 695–734 (2017).

[259] K. Sayevand and K. Pichaghchi, Successive approximation: a survey on stable manifold of fractional differential systems, *Fract. Calc. Appl. Anal.* **18**, pp. 621–641 (2015).

[260] G. R. Sell and Y. You, *Dynamics of Evolutionary Equations.* Springer-Verlag, New York (2002).

[261] E. Shishkina and S. Sitnik, *Transmutations, Singular and Fractional Differential Equations With Applications to Mathematical Physics.* Elsevier/Academic Press, Amsterdam (2020).

[262] J. L. Shomberg, Attractors for damped semilinear wave equations with singularly perturbed acoustic boundary conditions, *Electron. J. Differ. Equ.* **2018** (2018), paper ID 152.

[263] S. Siegmund and S. Piskarev, Approximations of stable manifolds in the vicinity of hyperbolic equilibrium points for fractional differential equations, *Nonlin. Dynam.* **95**, 1, pp. 685–697 (2019).

[264] S. Smale, Differentiable dynamical systems, *Bull. Am. Math. Soc.* **73**, pp. 747–817 (1967).

[265] P. E. Sobolevskii, The coercive solvability of difference equations, *Dokl. Akad. Nauk SSSR* **201**, pp. 1063–1066 (1971).

[266] P. E. Sobolevskii, Some properties of the solutions of differential equations in fractional spaces, *Tr. Nauch.-Issl. Inst. Mat. Voronezh. Univ.*, 74, pp. 68–76 (1975).

[267] P. E. Sobolevsky, The theory of semigroups and the stability of difference schemes, in *Operator Theory in Function Spaces.* Novosibirsk, pp. 304–337 (1977).

[268] A. Stuart, Perturbation theory for infinite-dimensional dynamical systems, in *Proc. Sixth SERC Summer School in Numerical Analysis, Leicester, UK, July 18–29, 1994.* Clarendon Press, Oxford, pp. 181–290 (1995).

[269] A. Stuart, Convergence and stability in the numerical approximation of dynamical systems, in *The State of the Art in Numerical Analysis.* Oxford Univ. Press, New York, pp. 145–169 (1997).

[270] A. M. Stuart and A. R. Humphries, *Numerical Analysis and Dynamical Systems.* Cambridge Univ. Press, Cambridge (1996).

[271] F. Stummel, Diskrete konvergenz linearer operatoren, in *Proc. Conf. "Linear Operators and Approximation,"* Oberwolfach Math. Res. Inst., 1971. Birkhäuser, Basel, pp. 196–216 (1972).

[272] F. Stummel, Perturbation of domains in elliptic boundary value problems, *Lect. Notes Math.* **503**, pp. 110–136 (1976).

[273] N. Tatar, The decay rate for a fractional differential equation, *J. Math. Anal. Appl.* **295**, pp. 303–314 (2004).

[274] A. A. Tateishi, H. V. Ribeiro and E. K. Lenzi, The role of fractional time-derivative operators on anomalous diffusion, *Front. Phys.* **5** (2017), paper ID 52.

[275] V. Thomée, Finite element methods for parabolic problems. some steps in the evolution, in *Finite Element Methods.* Marcel Dekker, New York, pp. 433–442 (1994).

[276] Y. Tomilov, A resolvent approach to stability of operator semigroups, *J. Operator Theory* **46**, pp. 63–98 (2001).

[277] D. K. Tran and V. T. Tran, Finite-time attractivity for semilinear fractional differential equations, *Res. Math.* **73**, 7, pp. 1–19 (2018).

[278] H. F. Trotter, Approximation of semigroups of operators, *Pac. J. Math.* **8**, pp. 887–919 (1958).

[279] H. F. Trotter, Approximation and perturbation of semigroups, in *Proc. Conf. "Linear Operators and Approximation," Oberwolfach Math. Res. Inst., 1974*. Birkhäuser, Basel, pp. 3–21 (1974).

[280] H. T. Tuan, S. Siegmund, D. T. Son and N. Cong, An instability theorem for nonlinear fractional differential systems, *Discr. Contin. Dynam. Syst. Ser. B* **22**, 8, pp. 3079–3090 (2017).

[281] E. H. Twizell, *Computational Methods for Partial Differential Equations*. Ellis Horwood, Chichester (1984).

[282] E. H. Twizell, *Numerical Methods, with Applications in the Biomedical Sciences*. Wiley, Chichester (1988).

[283] V. V. Uchaikin, *Fractional Derivatives for Physicists and Engineers*. Springer-Verlag, Berlin–Heidelberg (2013).

[284] T. Ushijima, Approximation theory for semigroups of linear operators and its application to approximation of wave equations, *Jpn. J. Math.* **1**, 1, pp. 185–224 (1975).

[285] G. Vainikko, *Funktionalanalysis der Diskretisierungsmethoden*. Teubner Verlag B.G., Leipzig (1976).

[286] G. Vainikko, Über die Konvergenz und Divergenz von Näherungsmethoden bei Eigenwertproblemen, *Math. Nachr.* **78**, pp. 145–164 (1977).

[287] G. Vainikko, Über Konvergenzbegriffe für lineare Operatoren in der numerischen Mathematik, *Math. Nachr.* **78**, pp. 165–183 (1977).

[288] G. Vainikko, Approximative methods for nonlinear equations (two approaches to the convergence problem), *Nonlin. Anal. Theory Meth. Appl.* **2**, pp. 647–687 (1978).

[289] G. Vainikko, Foundations of finite difference method for eigenvalue problems, in *Proc. Summer School "The Use of Finite Element Method and Finite Difference Method in Geophysics", Liblice, 1977*. Česk. Akad. Věd, Prague, pp. 173–192 (1978).

[290] G. Vainikko, Which functions are fractionally differentiable? *J. Anal. Appl.* **35**, pp. 465–487 (2016).

[291] G. Vainikko and S. Piskarev, Regularly compatible operators, *Izv. Vyssh. Ucheb. Zaved. Mat.*, 10, pp. 25–36 (1977).

[292] V. V. Vasiliev, S. G.Krein and S. I. Piskarev, Operator semigroups, cosine operator functions, and linear differential equations, *J. Sov. Math.* **54**, 4, pp. 1042–1129 (1991).

[293] V. V. Vasiliev and S. I. Piskarev, *Differential Equations in Banach Spaces. Semigroup Theory*. Moscow State Univ., Moscow (1996).

[294] V. V. Vasiliev and S. I. Piskarev, *Differential Equations in Abstract Spaces. Bibliographic Index*. Moscow State Univ., Moscow (2001).

[295] V. V. Vasiliev and S. I. Piskarev, Differential equations in Banach spaces. ii. cosine-operator functions, *J. Math. Sci.* **122**, 2, pp. 3055–3174 (2004).

[296] V. V. Vasiliev, S. I. Piskarev and N. Y. Selivanova, Integrated semigroups and c-semigroups and their applications, *J. Math. Sci.* **230**, 4, pp. 513–646 (2018).

[297] V. Vijayakumar, R. Murugesu, R. Poongodi, and S. Dhanalakshmi, Controllability of second-order impulsive nonlocal Cauchy problem via measure of noncompactness, *Mediterr. J. Math.* **14**, 1 (2017), paper ID 3.

[298] M. I. Vishik and V. V. Chepyzhov, Approximation of trajectories lying on a global attractor of a hyperbolic equation with exterior force rapidly oscillating in time, *Sb. Math.* **194**, 9, pp. 1273–1300 (2003).

[299] M. I. Vishik and V. V. Chepyzhov, Trajectory attractors of equations of mathematical physics, *Russ. Math. Surv.* **66**, 4, pp. 637–731 (2011).

[300] J. Voigt, On the convex compactness property for the strong operator topology, *Note Mat.* **12**, pp. 259–269 (1992).

[301] I. I. Vrabie, A class of semilinear delay differential equations with nonlocal initial conditions, *Dynam. Partial Differ. Equ.* **15**, 1, pp. 45–60 (2018).

[302] Vu Quoc Phong, On the exponential stability and dichotomy of c_0-semigroups, *Stud. Math.* **132**, 2, pp. 141–149 (1999).

[303] Vu Quoc Phong, The spectral radius, hyperbolic operators and lyapunov's theorem, in *Evolution Equations and Their Applications in Physical and Life Sciences*. Marcel Dekker, New York, pp. 187–194 (2001).

[304] Vu Quoc Phong, A new proof and generalizations of gearhart's theorem, *Proc. Am. Math. Soc.* **135**, 7, pp. 2065–2072 (2007).

[305] Vu Quoc Phong and E. Schuler, The operator equation $ax - xb = c$, admissibility, and asymptotic behavior of differential equations, *J. Differ. Equ.* **145**, 2, pp. 394–419 (1998).

[306] R. N. Wang, D. H. Chen and T. J. Xiao, Abstract fractional Cauchy problems with almost sectorial operators, *J. Differ. Equ.* **252**, pp. 202–235 (2012).

[307] R. N. Wang, T. J. Xiao and J. Liang, A note on the fractional Cauchy problems with nonlocal initial conditions, *Appl. Math. Lett.* **24**, pp. 1435–1442 (2011).

[308] Y. Wang, M. Hu and Y. Qin, Upper semicontinuity of pullback attractors for a nonautonomous damped wave equation, *Boundary-Value Probl.* **56**, pp. 1–19 (2021).

[309] Z. Xia, D. Wang, C.-F. Wen and J.-C. Yao, Pseudo asymptotically periodic mild solutions of semilinear functional integro-differential equations in Banach spaces, *Math. Meth. Appl. Sci.* **40**, 18, pp. 7333–7355 (2017).

[310] J.-Z. Xiao, Z.-Y. Wang and J. Liu, Hausdorff product measures and c^1-solution sets of abstract semilinear functional differential inclusions, *Topol. Meth. Nonlin. Anal.* **49**, 1, pp. 273–298 (2017).

[311] L. Ya-Ning and S. Hong-Rui, Regularity of mild solutions to fractional Cauchy problems with Riemann–Liouville fractional derivative, *Electron. J. Differ. Equ.* **2014** (2014), paper ID 184.

[312] C. Yong-Kui and J. Zhao, Some new asymptotic properties on solutions to fractional evolution equations in Banach spaces, *Appl. Anal.* (2021).

[313] K. Yosida, *Functional Analysis*. Springer-Verlag, Berlin (1965).

[314] J. Yu and Z. Wang, Existence of mild solutions for nonlocal impulsive evolution equations with noncompact semigroups, *J. Nanchang Univ. Nat. Sci.* **40**, 4, pp. 319–323 (2016).

[315] H. Zhang, W. Zhang and Q. Hu, Global existence and blow-up of solution for the semilinear wave equation with interior and boundary source terms, *Boundary-Value Probl.* **2019** (2019), paper ID 18.

[316] J. Zhang, C. Zhong and B. You, The existence of multiple equilibrium points in global attractors for some symmetric dynamical systems, ii, *Nonlin. Anal. Real World Appl.* **36**, pp. 44–55 (2017).

[317] Q. Zhang and G. Li, Global attractors of strongly damped wave equations with critical nonlinearities, *Acta Math. Appl. Sin.* **40**, 2, pp. 192–203 (2017).

[318] X. Zhao, Fourier spectral approximation to global attractor for 2d convective Cahn–Hilliard equation, *Bull. Malays. Math. Sci. Soc.* (2) **41**, 2, pp. 1119–1138 (2018).

Index

Printed in the United States
by Baker & Taylor Publisher Services